Computational Modeling and Data Analysis in COVID-19 Research

Emerging Trends in Biomedical Technologies and Health informatics Series

Series Editors:
Subhendu Kumar Pani
Orissa Engineering College, Bhubaneswar, Orissa, India

Sujata Dash
North Orissa University, Baripada, India

Sunil Vadera
University of Salford, Salford, UK

Everyday Technologies in Healthcare
Chhabi Rani Panigrahi, Bibudhendu Pati, Mamata Rath, Rajkumar Buyya

For more information about this series, please visit: https://www.routledge.com/Emerging-Trends-in-Biomedical-Technologies-and-Health-informatics-series/book-series/ETBTHI

Computational Modeling and Data Analysis in COVID-19 Research

Edited by Chhabi Rani Panigrahi,
Bibudhendu Pati, Mamata Rath
and Rajkumar Buyya

Boca Raton London New York

CRC Press is an imprint of the
Taylor & Francis Group, an **informa** business

MATLAB® is a trademark of The MathWorks, Inc. and is used with permission. The MathWorks does not warrant the accuracy of the text or exercises in this book. This book's use or discussion of MATLAB® software or related products does not constitute endorsement or sponsorship by The MathWorks of a particular pedagogical approach or particular use of the MATLAB® software.

First edition published 2021
by CRC Press
6000 Broken Sound Parkway NW, Suite 300, Boca Raton, FL 33487-2742

and by CRC Press
2 Park Square, Milton Park, Abingdon, Oxon, OX14 4RN

© 2021 Taylor & Francis Group, LLC

CRC Press is an imprint of Taylor & Francis Group, LLC

The right of Chhabi Rani Panigrahi, Bibudhendu Pati, Mamata Rath & Rajkumar Buyya to be identified as the authors of the editorial material, and of the authors for their individual chapters, has been asserted in accordance with sections 77 and 78 of the Copyright, Designs and Patents Act 1988.

Reasonable efforts have been made to publish reliable data and information, but the author and publisher cannot assume responsibility for the validity of all materials or the consequences of their use. The authors and publishers have attempted to trace the copyright holders of all material reproduced in this publication and apologize to copyright holders if permission to publish in this form has not been obtained. If any copyright material has not been acknowledged please write and let us know so we may rectify in any future reprint.

Except as permitted under U.S. Copyright Law, no part of this book may be reprinted, reproduced, transmitted, or utilized in any form by any electronic, mechanical, or other means, now known or hereafter invented, including photocopying, microfilming, and recording, or in any information storage or retrieval system, without written permission from the publishers.

For permission to photocopy or use material electronically from this work, access www.copyright.com or contact the Copyright Clearance Center, Inc. (CCC), 222 Rosewood Drive, Danvers, MA 01923, 978-750-8400. For works that are not available on CCC please contact mpkbookspermissions@tandf.co.uk

Trademark notice: Product or corporate names may be trademarks or registered trademarks and are used only for identification and explanation without intent to infringe.

Library of Congress Cataloging-in-Publication Data

Names: Rani Panigrahi, Chhabi, editor. | Pati, Bibudhendu, editor. | Rath, Mamata, editor. | Buyya, Rajkumar, 1970- editor.
Title: Computational modeling and data analysis in COVID-19 research / edited by Chhabi Rani Panigrahi, Bibudhendu Pati, Mamata Rath & Rajkumar Buyya.
Description: First edition. | Boca Raton : CRC Press, 2021. | Includes bibliographical references and index. | Summary: "This book covers recent research on the COVID-19 pandemic. It includes the analysis, implementation, usage and proposed ideas and models with architecture to handle the COVID-19 outbreak. Using advanced technologies such as AI and ML techniques for data analysis, this book will be helpful to mitigate exposure and ensure public health"-- Provided by publisher.
Identifiers: LCCN 2020054427 (print) | LCCN 2020054428 (ebook) | ISBN 9780367680367 (hbk) | ISBN 9781003137481 (ebk)
Subjects: MESH: Coronavirus Infections | Severe Acute Respiratory Syndrome | Computer Simulation | Models, Biological | Data Analysis | Research Design
Classification: LCC RA644.C67 (print) | LCC RA644.C67 (ebook) | NLM WC 505 | DDC 616.2/4140072--dc23
LC record available at https://lccn.loc.gov/2020054427
LC ebook record available at https://lccn.loc.gov/2020054428

ISBN: 978-0-367-68036-7 (hbk)
ISBN: 978-0-367-68428-0 (pbk)
ISBN: 978-1-003-13748-1 (ebk)

Typeset in Times
by Deanta Global Publishing Services, Chennai, India

Access the [companion website/Support Material]: [insert comp website/ Support Material URL]

Contents

Preface vii
Editors ix
Contributors xi

Chapter 1   Machine Learning Implementations in COVID-19 1

Kabita Kumari, S.K. Pahuja, and Sanjeev Kumar

Chapter 2   Analysis of COVID-19 Data Using Consensus
Clustering Technique 17

Arko Banerjee, Sunandana Mukherjee, Chhabi Rani Panigrahi, Bibudhendu Pati, and Rajib Mall

Chapter 3   *MoBMGAN*: Modified GAN-Based Transfer Learning for Automatic Detection of COVID-19 Cases Using Chest X-ray Images 29

Rajashree Nayak, Bunil Ku. Balabantaray and Dipti Patra

Chapter 4   Application and Progress of Drone Technology in the COVID-19 Pandemic: A Comprehensive Review.............................. 47

Vasundhara Saraf, Lipsita Senapati and Tripti Swarnkar

Chapter 5   Smart War on COVID-19 and Global Pandemics:
Integrated AI and Blockchain Ecosystem 67

Anil D. Pathak, Debasis Saran, Sibani Mishra, Madapathi Hitesh, Sivaiah Bathula, and Kisor K. Sahu

Chapter 6   Machine Learning-Based Text Mining in Social Media for COVID-19 95

Tajinder Singh and Madhu Kumari

Chapter 7   Containing the Spread of COVID-19 with IoT: A Visual Tracing Approach 127

Pallav Kumar Deb, Sudip Misra, Anandarup Mukherjee, and Aritra Bandyopadhyay

Chapter 8 Crowd-Sourced Centralized Thermal Imaging for Isolation and Quarantine 145

Sudershan Kumar, Prabuddha Sinha, and Sujata Pal

Chapter 9 Blockchain Technology for Limiting the Impact of Pandemic: Challenges and Prospects 165

Suchismita Swain, Oyekola Peter, Ramasamy Adimuthu, and Kamalakanta Muduli

Chapter 10 A Study on Mathematical and Computational Models in the Context of COVID-19 187

Dr. Meera Joshi

Chapter 11 A Detailed Study on AI-Based Diagnosis of Novel Coronavirus from Radiograph Images 209

Malaya Kumar Nath and Aniruddha Kanhe

Chapter 12 Data Analytics for COVID-19 231

Shreyas Mishra

Index 255

Preface

This book covers recent researches on the COVID-19 pandemic. It includes the implementation, analysis, usage, and proposed ideas and models with architecture to handle the COVID-19 outbreak. The chapters are written by leading international researchers from industry, academia, government, and private research institutions, and this book offers a broad view of important developments in COVID-19 research. This book presents:

- An extensive survey on machine learning implementations in COVID-19.
- Recent research results on the proposed topic using emerging technologies such as artificial intelligence, machine learning, data analytics, drone technologies, image processing, IoT, and cloud computing technology.
- A study of mathematical and computational models in the context of COVID-19.

This book will help to expand the reader's knowledge in the application of artificial intelligence to handle the COVID-19 pandemic and to continue their further research in this area.

MATLAB® is a registered trademark of The MathWorks, Inc. For product information, please contact:

The MathWorks, Inc.
3 Apple Hill Drive
Natick, MA 01760-2098 USA
Tel: 508 647 7000
Fax: 508-647-7001
E-mail: info@mathworks.com
Web: www.mathworks.com

Editors

Chhabi Rani Panigrahi is Assistant Professor in the Department of Computer Science at Rama Devi Women's University, Bhubaneswar, India. Prior to this, she was working as Assistant Professor at Central University of Rajasthan, India. She earned her PhD in the Department of Computer Science and Engineering, Indian Institute of Technology Kharagpur, India. Her areas of research interests include software testing, mobile cloud computing, and machine learning. She holds 19 years of teaching and research experience. She has published several international journals, conference papers, and books. She is a Life Member of the Indian Society for Technical Education (ISTE) and a Member of the IEEE and Computer Society of India (CSI). She has also served as Guest Editor of many reputed journals of Inderscience publications. She was the Organizing Chair of ICACIE 2016, ICACIE 2017, ICACIE 2018, ICACIE 2019, ICACIE 2020, and WiE Chair of IEEE ANTS 2017 and IEEE ANTS 2018. She is also associated as a reviewer in many journals.

Bibudhendu Pati is Associate Professor and Head of the Department of Computer Science at Rama Devi Women's University, Bhubaneswar, India. He has around 21 years of experience in teaching and research. His areas of research interests include wireless sensor networks, cloud computing, big data, Internet of Things, and advanced network technologies. He earned his PhD from IIT Kharagpur, India. He is a Life Member of the Indian Society for Technical Education, Computer Society of India, and Senior Member of IEEE. He has several papers published in reputed journals, conference proceedings, and books of international repute. He has also served as Guest Editor of IJCNDS and IJCSE journals. He was the General Chair of ICACIE 2016, ICACIE 2018, ICACIE 2019, ICACIE 2020, and IEEE ANTS 2017 international conferences. He has been involved in many professional and editorial activities.

Mamata Rath earned her PhD in Computer Science from Siksha 'O' Anusandhan University, Odisha, India. She has completed her MTech in Computer Science from Biju Patnaik University of Technology, Odisha, India. Currently, she is working as Assistant Professor, Information Technology, at Birla Global University, Bhubaneswar, India. Her research interests include information systems, mobile networks, Internet of Things, and computer security. She has a good number of research publications in reputed journals with Scopus and SCI indexing on topics like cognitive radio network, ubiquitous computing, wireless network security, information systems, IoT, big data, and social networking. She has published various book chapters as well as presented papers at international conferences. She reviews research articles for reputed journals and international conferences of repute.

Rajkumar Buyya is a Redmond Barry Distinguished Professor and Director of the Cloud Computing and Distributed Systems (CLOUDS) Laboratory at the University of Melbourne, Australia. He is also serving as the founding CEO of Manjrasoft, a spin-off company of the university, commercializing its innovations in cloud computing. He served as a Future Fellow of the Australian Research Council during 2012–2016. He has authored over 625 publications and 7 textbooks including *Mastering Cloud Computing* published by McGraw Hill, China Machine Press, and Morgan Kaufmann for Indian, Chinese, and international markets, respectively. He has also edited several books including *Cloud Computing: Principles and Paradigms*" (Wiley Press, USA, February 2011). He is one of the most highly cited authors in computer science and software engineering worldwide (h-index = 123, g-index = 271, 79,000+ citations). "A Scientometric Analysis of Cloud Computing Literature" by German scientists ranked Dr. Buyya as the World's Top-Cited (#1) Author and the World's Most-Productive (#1) Author in Cloud Computing. Dr. Buyya has been recognized as a Web of Science Highly Cited Researcher for three consecutive years since 2016, a Fellow of IEEE, and Scopus Researcher of the Year 2017 with Excellence in Innovative Research Award by Elsevier for his outstanding contributions to cloud computing. Software technologies for grid and cloud computing developed under Dr. Buyya's leadership have gained rapid acceptance and are in use at several academic institutions and commercial enterprises in 40 countries around the world. Dr. Buyya has led the establishment and development of key community activities, including serving as foundation Chair of the IEEE Technical Committee on Scalable Computing and five IEEE/ACM conferences. These contributions and international research leadership of Dr. Buyya are recognized through the 2009 IEEE Medal for Excellence in Scalable Computing from the IEEE Computer Society TCSC. Manjrasoft's Aneka Cloud technology developed under his leadership has received 2010 Frost & Sullivan New Product Innovation Award. Recently, Dr. Buyya received the Mahatma Gandhi Award, along with gold medals for his outstanding and extraordinary achievements in information technology field and services rendered to promote greater friendship and India–International cooperation. He served as the founding Editor-in-Chief of the IEEE Transactions on Cloud Computing. He is currently serving as Co-Editor-in-Chief for the journal *Software: Practice and Experience*, which was established around 50 years ago. For further information on Dr. Buyya, please visit his cyberhome: www.buyya.com

Contributors

Ramasamy Adimuthu
Department of Business Studies
Papua New Guinea University of Technology
Lae, Morobe, Papua New Guinea

Bunil Ku. Balabantaray
National Institute of Technology Meghalaya
Meghalaya, India

Aritra Bandyopadhyay
Manipal Institute of Technology
Karnataka, India

Arko Banerjee
College of Engineering and Management
Kolaghat, India

Sivaiah Bathula
School of Minerals, Metallurgical and Materials Engineering
Indian Institute of Technology Bhubaneswar
Bhubaneswar, India

Pallav Kumar Deb
Indian Institute of Technology Kharagpur
Kharagpur, India

Madapathi Hitesh
School of Minerals, Metallurgical and Materials Engineering
Indian Institute of Technology Bhubaneswar
Bhubaneswar, India

and

School of Electrical Sciences
Indian Institute of Technology Bhubaneswar
Bhubaneswar, India

Meera Joshi
Department of Mathematics
Aurora Degree & PG College
Hyderabad, India

Aniruddha Kanhe
Department of ECE
National Institute of Technology Puducherry
Karaikal, India

Sanjeev Kumar
Academy of Scientific and Innovative Research (AcSIR)
CSIR-Central Scientific Instruments Organisation (CSIR-CSIO)
Chandigarh, India

Sudershan Kumar
Department of CSE
Indian Institute of Technology Ropar
Ropar, India

Kabita Kumari
ICE Department
Dr. B. R. Ambedkar National Institute
  of Technology
Jalandhar, India

Madhu Kumari
Indian Institute of Management
  Amritsar
Amritsar, India

Rajib Mall
Indian Institute of Technology
Kharagpur, India

Shreyas Mishra
National Institute of Technology
Rourkela, India

Sibani Mishra
School of Minerals, Metallurgical and
  Materials Engineering
Indian Institute of Technology
  Bhubaneswar
Bhubaneswar, India

Sudip Misra
Indian Institute of Technology
Kharagpur, India

Kamalakanta Muduli
Department of Mechanical Engineering
Papua New Guinea University of
  Technology
Lae, Morobe, Papua New Guinea

Anandarup Mukherjee
University of Cambridge
Cambridge, England

Sunandana Mukherjee
Sammilani Mahavidyalaya
Kolkata, India

Malaya Kumar Nath
Department of ECE
NIT Puducherry
Karaikal, India

Rajashree Nayak
National Institute of Technology
  Meghalaya
Meghalaya, India

S.K. Pahuja
ICE Department
Dr. B. R. Ambedkar National Institute
  of Technology
Jalandhar, India

Sujata Pal
Department of CSE
Indian Institute of Technology Ropar
Ropar, India

Chhabi Rani Panigrahi
Department of Computer Science
Rama Devi Women's University
Bhubaneswar, India

Anil D. Pathak
ElectroGati Technologies Pvt. Ltd.
Maharashtra, India

Bibudhendu Pati
Department of Computer Science
Rama Devi Women's University
Bhubaneswar, India

Dipti Patra
National Institute of Technology
  Rourkela
Rourkela, India

Oyekola Peter
Department of Mechanical Engineering
Papua New Guinea University of
  Technology
Lae, Morobe, Papua New Guinea

Contributors

Kisor K. Sahu
School of Minerals, Metallurgical and Materials Engineering
Indian Institute of Technology Bhubaneswar
Bhubaneswar, India

and

Virtual and Augmented Reality Centre of Excellence (VARCOE)
Indian Institute of Technology Bhubaneswar
Bhubaneswar, India

Vasundhara Saraf
Department of Mechanical Engineering
Veer Surendra Sai University of Technology
Burla, India

Debasis Saran
School of Minerals, Metallurgical and Materials Engineering
Indian Institute of Technology Bhubaneswar
Bhubaneswar, India

Lipsita Senapati
Application Development Service Management (RBEI/BSA)
Robert Bosch Engineering and Business Solutions Private Limited
Bengaluru, India

Prabuddha Sinha
Department of CSE
Indian Institute of Technology Ropar
Ropar, India

Tajinder Singh
Indian Institute of Management
Kashipur, India

Suchismita Swain
Department of Mechanical Engineering
Biju Patnaik University of Technology
Rourkela, India

Tripti Swarnkar
Department of Computer Application
Siksha 'O' Anusandhan (Deemed to be University)
Bhubaneswar, India

1 Machine Learning Implementations in COVID-19

Kabita Kumari, S.K. Pahuja, and Sanjeev Kumar

CONTENTS

1.1 Introduction 1
    1.1.1 A Brief History of Coronaviruses...2
    1.1.2 Current Scenario of COVID-19..3
    1.1.3 Symptoms, Diagnosis, and Preventions Steps for COVID-19..............3
1.2 Machine Learning in COVID-19 Outbreak 7
    1.2.1 Application of ML in the War on COVID-19.......................................7
    1.2.2 Machine Learning in COVID-19 Future Forecasting 8
    1.2.3 Machine Learning as a Diagnostic Tool for COVID-19 9
    1.2.4 COVID Patient's Health Prediction Using Machine Learning 11
    1.2.5 Machine Learning in Survival Analysis and Discharge Time
           Prediction of COVID-19 Patients 11
    1.2.6 Machine Learning in Contact Tracing of COVID-19 12
1.3 Conclusion 12
References... 13

1.1 INTRODUCTION

There is currently a coronavirus emergency (COVID-19) that has spread worldwide, threatening lives and disrupting daily life. A number of pandemics have attacked humankind in world history. Coronaviruses (CoVs), as the name hints, are a large group of viruses that cause ailments which vary from the usual cold and cough to more severe infections like Middle East Respiratory Syndrome (MERS-CoV) and Severe Acute Respiratory Syndrome (SARS-CoV). COVID-19 is a zoonotic infection brought about by germs that spread between animals and humans. Some common symptoms of COVID-19 are fever, cough, tiredness, and breathing challenges. However, it can also cause pneumonia, intense respiratory disorders, and kidney damage in more severe cases. COVID-19 has affected 213 countries and territories around the globe. On December 12, 2019, COVID was first recognized in the city of Wuhan, Hubei region, China. On February 11, 2020, the World Health Organization (WHO) officially declared COVID-19 as a global crisis brought about by SARS-CoV-2. SARS-CoV-2 has infected more than 94 million individuals and caused 2,050,857 deaths to date, with those numbers ever-expanding. It has affected almost all aspects of life in many nations and influenced our day-to-day life, families, and societies. To date, the WHO, its co-working physicians, and numerous public experts worldwide are battling this pandemic. In this global emergency, platforms are needed that accelerate the demand for diagnosis and treatment of COVID-19. Machine learning (ML) is one such platform that has substantiated itself as an important field of study during the last decade by solving numerous complex and refined issues (Jamshidi et al. 2020). It has been applied in almost every field, including medical services, autonomous vehicles (AV), business applications, natural language processing (NLP), smart robots, gaming, atmospheric modeling, voice recognition, and image preparation. In this chapter, we are concentrating on the recent ongoing applications and developments in machine learning that have helped with coronavirus treatment, medication processes, its prediction, contact tracing, and so on. Machine learning techniques have additionally reduced the workload of humans during the pandemic.

1.1.1 A Brief History of Coronaviruses

The flare-ups began in China and were brought about by viruses that originated in bats. Coronaviruses (CoVs) are in the subfamily Orthocoronavirinae in the family Coronaviridae, in the order Nidovirales. The name "coronavirus" is derived from the Latin word *corona*, which means "crown," i.e., a crown-like virus particle (Figure 1.1). CoVs are a large group of viruses, generally circular particles with unique surface projections and consist of a positive-sense genome, single-strand RNA (ribonucleic acid), whose size varies from 27 to 32 Kb, and is the second-largest RNA virus genome (Fan et al. 2019).

Contrasted with other RNA infections, the extended genome size of CoVs was found to be related to expanded replication constancy, after getting qualities

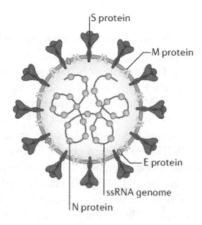

FIGURE 1.1 Structure of SARS-CoVs. (Nishiga et al. 2020.)

encoding RNA-handling enzymes resulting from genome changes brought about by recombination, quality exchange, and quality inclusion or cancellation among CoVs. CoVs can cause disease in an assortment of domestic and wild creatures just as in people, where α- and β-CoVs fundamentally affect warm-blooded creatures and γ- and δ-CoVs mostly affect winged animals. There are two highly infectious viruses, β-coronaviruses SARS (Severe Acute Respiratory Syndrome), which was first identified in China in 2003, and MERS (Middle East Respiratory Syndrome), which was first identified in Saudi Arabia in 2012, which have caused pandemics among humankind (Zaki et al. 2012). The origin of the coronaviruses was in China, and afterward it spread to different parts of the world; SARS-CoV contaminated around 8,000 people with a general mortality of 10% during the 2002–2003 pandemic (Fan et al. 2019). Some of the species, reservoir hosts of coronaviruses, and their presence reported in China are listed in Table 1.1, classified by the International Committee of Taxonomy of Viruses (ICTV).

1.1.2 Current Scenario of COVID-19

By the end of 2019, the news showed a pandemic of pneumonia with some cases in Wuhan, China's seafood and live animal market. Initially, a few more points were revealed around December 8, reported by China to the World Health Organization (WHO). On December 12, 2019, COVID was first recognized in Wuhan city, Hubei region, China. On January 13, 2020, the first case of the novel coronavirus was confirmed outside China in Thailand. On January 30, 2020, the World Health Organization (WHO) made the official declaration of COVID-19 as a Public Health Emergency of International Concern (PHEIC) (WHO 2020). The novel COVID-19 has spread to more than 227 countries and territories with a total of 28,637,952 confirmed cases, including 917,417 deaths by September 13, 2020 (Rampal and Seng 2020; WHO 2020). The 227 countries' top 9 governments with a cumulative number of confirmed cases and deaths per the WHO report are shown in Table 1.2, and on a graph (Figure 1.2).

Coronaviruses are a large group of viruses that consist of a core of genetic material encompassed by a lipid wrap with protein spikes. Various sorts of COVIDs cause disease in both animals and humans. Coronaviruses can cause respiratory infections in humans ranging from the common cold to more severe illnesses such as SARS and COVID-19. Coronaviruses are transmitted from animals and birds to humans via a mutation in the virus and increased contact between humans and animals. These viruses are further spread from person to person in the form of droplets. For example, when an infected person coughs, sneezes, or talks, these droplets spread; also, when these droplets fall on an object and a healthy person touches that object and then touches their eyes, nose, or mouth, they may contract the infection.

1.1.3 Symptoms, Diagnosis, and Preventions Steps for COVID-19

The exposure period for the appearance of the symptoms of COVID-19 is 5–6 days. Still, sometimes it can range from 10 to 14 days, and it can run from very mild to

TABLE 1.1
Shows the Species, Reservoir Host of Coronaviruses, and Its Presence Reported in China (Fan et al. 2019)

S. No.	Coronavirus Species	Abbreviations	Human	Bats	Other Animals	Reported in China
1	Human coronavirus 229E	HCoV-229E	Yes	—	—	Yes
2	Mink coronavirus 1	MCoV	—	—	Yes (mink)	No
3	Porcine epidemic diarrhea virus	PEDV	—	—	Yes (pig)	Yes
4	*Rhinolophus* bat coronavirus HKU2 (SADS)	BtRhCoV-HKU2	—	Yes	Yes	Yes
5	Human coronavirus NL63	HCoV-NL63	Yes	—	—	Yes
6	Alphacoronavirus 1 (Transmissible gastroenteritis virus)	TGEV	—	—	Yes (pig)	Yes
7	Human coronavirus HKU1	HCoV-HKU1	Yes	—	—	Yes
8	Murine coronavirus (Murine hepatitis coronavirus)	MHV	—	—	Yes (mouse)	No
9	Middle East respiratory syndrome-related coronavirus	MERSr-CoV	Yes	Yes	—	Yes
10	Severe acute respiratory syndrome-related Coronavirus	SARSr-CoV	Yes	Yes	—	Yes
11	Betacoronavirus 1 (Human coronavirus OC43)	HCoV-OC43	Yes	—	—	Yes
12	Wigeon coronavirus HKU20	WiCoV-HKU20	—	—	Yes (bird)	Yes
13	Avian infectious bronchitis virus	IBV	—	—	Yes (bird)	Yes
14	White-eye coronavirus HKU16	WECoV-HKU13	—	—	Yes (bird)	Yes
15	Common moorhen coronavirus HKU21	CMCoV-HKU21	—	—	Yes (bird)	Yes

TABLE 1.2
Top 9 Countries with Cumulative Confirmed and Cumulative Death Cases of COVID-19 (September 2020, WHO 2020)

Countries	Total Cumulative Confirmed Cases	Total Cumulative Deaths
United States	63,86,832	1,91,809
India	47,54,356	78,586
Brazil	42,82,164	1,30,396
Russia	10,62,811	18,578
Peru	7,16,670	30,470
Colombia	7,02,088	22,518
Mexico	6,58,299	70,183
South Africa	6,48,214	15,427
Spain	5,66,326	29,747

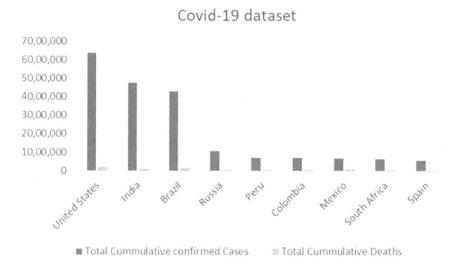

FIGURE 1.2 Analysis of COVID-19 data for the top 9 highly affected countries.

severe symptoms. Some people might not develop symptoms because they possess higher immunity levels. Some of the common symptoms are listed below with a graphical representation shown in Figure 1.3.

1. Fever
2. Fatigue
3. Respiratory problems like cough, sore throat, and difficulty in breathing
4. Loss of taste or smell and skin rashes in some people
5. Pneumonia, organ failure, and sometimes death in severe cases

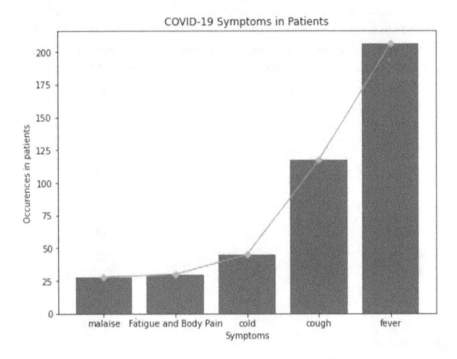

FIGURE 1.3 Symptoms in patients. (Iwendi et al. 2020.)

Almost 80% of infected people quickly recover from this disease without requiring any special treatment, but older people and people with medical issues such as diabetes, high blood pressure, heart disease, cancer, and chronic respiratory disease are at a higher risk of suffering serious complications. There are some diagnostic processes being used to detect COVID-19, namely Reverse Transcriptase Polymerase Chain Reaction (RT-PCR) tests, which identify the virus present in the DNA of the patient, and a blood test, which shows the antibody against the virus, providing information on previous infection. By the continuous efforts of the researchers to develop safe and effective vaccines against COVID-19, finally some of the vaccines have been developed. WHO have launched COVAX platform to reinforce research and development of COVID-19 vaccine. Pfizer is the first vaccine authorized for emergency use by a stringent regulatory authority. However, some of the preventive steps that will help in reducing the number of COVID-19 cases are as follows:

1. Mouth and nose must be covered with some tissue while sneezing and coughing and these tissues must be thrown in a dustbin after use
2. Wash hands regularly with soap or sanitize hands with a hand rub
3. Maintain a social distance of at least 1 meter and avoid close contact
4. Use masks and personal protective equipment (PPE) while going outside

5. Stay at home if you are feeling unwell
6. In cases of fever, cough, or difficulty in breathing, immediately seek medical attention

1.2 MACHINE LEARNING IN COVID-19 OUTBREAK

To control coronavirus's spread and save people's lives, countries' governments called for a complete lockdown. The lives of the people changed drastically, and the economic scale of most countries also fell. A number of researchers, scientists, doctors, and clinicians from all over the world are putting in all their efforts to tackle this problem as there is a need for early detection, fast diagnosis, and screening systems to save more lives. When there is a considerable amount of data, and complexity in health care rises, machine learning (ML) is highly useful to cope with this kind of problem (Lalmuanawma, Hussain, and Chhakchhuak 2020). Machine learning is a statistical tool for fitting models to data and "learning" via preparing models with data, and performance of ML-based algorithms and systems heavily depend on the representative features. ML is a subset of artificial intelligence (AI); so with the flow chart shown in Figure 1.4, we can describe the different application areas of ML-based methods to overcome the challenges of COVID-19. The first requirement is preparing a dataset for data mining. Datasets include medical information, clinical reports, images, or any form of data that can be transformed into useful data, i.e., data that can be understood by the machine. The data is reformatted and corrected before being processed and analyzed, and the collection and analysis of data such as physical and clinical data result in big data. Afterward, at this phase, human intercession, as a part of ML techniques, happens, and specialists examine and investigate the information to extract the data with the best structures, examples, and highlights. Deep learning (DL) methods can be used when an enormous dataset challenges ML, as DL is a subset of ML. ML algorithms are usually dependent on structured data.

Unlike supervised learning, which is the task of learning a function which maps an input to an output based on example input-output pairs, unsupervised learning is marked by minimum involvement of human supervision. It could be described as a sort of machine learning, searching for undetected patterns in a dataset where no prior labels exist. Other conventional practices include medical care suppliers, allopathic medicine, biomedicine, etc. (Jamshidi et al. 2020). There is much evidence in the medical field that proves machine learning algorithms provide effective models to solve problems, and many researchers and scientists believe in it. They are involved with this technique to face the challenges posed by the coronavirus (Chaurasia and Pal 2020). To date, various ML applications and methods are being used, and much research on ML application is still ongoing.

1.2.1 Application of ML in the War on COVID-19

This chapter presents a comprehensive study on some of the machine learning techniques employed to solve the challenges of the COVID-19 outbreak and also discusses the performance and advantages and disadvantages of the methods used.

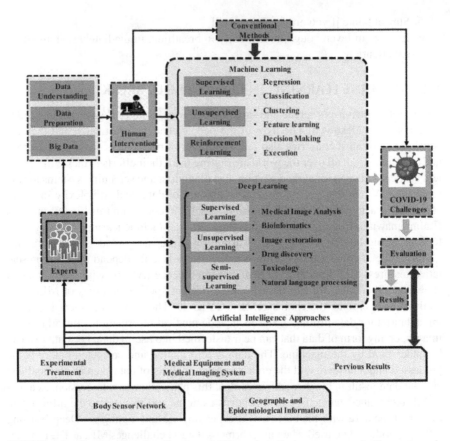

FIGURE 1.4 The process of application of AI-based methods to conquer challenges associated with COVID-19. (Jamshidi et al. 2020.)

1.2.2 Machine Learning in COVID-19 Future Forecasting

Thousands of people are becoming infected by COVID-19 every day, and one of the most dangerous aspects of this disease is that a person carries the virus for several days without showing any symptoms. Under such circumstances, a further prediction method will be beneficial in recognizing and controlling this disease. There are different prediction methods for various conditions using machine learning like breast cancer prediction, cardiovascular disease prediction, and coronary artery disease prediction (Lapuerta, Azen, and LaBree 1995; Uno et al. 2011; Asri et al. 2016). This study supervised learning techniques to forecast the number of positive cases and death cases and the number of recoveries (Rustam et al. 2020). This type of estimating has been thought of as a relapse issue in this examination, so the investigation is based on four different regression models: linear regression (LR), least absolute shrinkage and selection operator (LASSO), support vector machine (SVM), and exponential smoothing (ES). The learning models have been prepared to utilize the COVID-19 patient stat dataset given by Johns Hopkins University. This dataset has been preprocessed and isolated into two subsets: preparing set (85% records) and testing set (15% records).

The performance assessment is carried out in terms of substantial measures, including R-squared score (R2 score), adjusted R-squared (R2 adjusted), mean square error (MSE), mean absolute error (MAE), and root mean square error (RMSE). To quantify efficiency, models were trained from the dataset from January 20, 2020, to March 22, 2020, at an interval of 10–15 days, as the prediction interval indicates a probabilistic upper and lower leap in estimating the outcome. Among all four models, the ES model performs best in the forecasting domain, even with a limited number of datasets, followed by LR, LASSO, and SVM. As per experimental outcomes, the models provide correct forecasting of the current scenario, which will help governments understand upcoming situations and take preventive actions against COVID-19.

In a study by Chakraborty and Ghosh (2020), data-driven analysis for real-time forecasting and risk assessment on COVID-19 cases was done. The forecasting mostly shows an oscillating behavior in an interval of 10 days and reflects the impact of social distancing measures by government which stabilizes the epidemic. A simplified ML approach, i.e., optimal regression tree model, is used to assess the possibility of COVID-19 and to make associations with fatality rates using 7 key parameters for 50 positively affected nations. A hybrid methodology combining autoregressive integrated moving average (ARIMA) and wavelet based forecasting (WBF) models is proposed for the prediction of the disease. An RT (regression tree) based risk assessment with case fatality rate (CFR) for 50 countries, provides a better visual representation and is easily interpretable. Hence, forecasting of the future outbreak for different countries will benefit the compelling portion of medical services assets and will go about as an early cautioning framework for government policymakers.

1.2.3 Machine Learning as a Diagnostic Tool for COVID-19

The early detection and fast diagnosis of any disease or infection help reduce the spread of viruses and save more lives. Machine learning and artificial intelligence help diagnose infectious cases by employing medical imaging processes like computed tomography (CT) and magnetic resonance imaging (MRI) (Vaishya et al. 2020). In the following study (Elaziz et al. 2020), a new machine learning method was developed for the diagnosis of COVID-19 using images of chest X-ray, as COVID-19 highly affects our respiratory system. Machine learning has been used in the field of image processing, such as image analysis, classification, and segmentation (Ke et al. 2019; Chouhan et al. 2020; Yang et al. 2017). The X-ray images are divided into two classes: one with COVID-19 patients and one with non-COVID-19 patients. A new Fractional Multichannel Exponent Moments (FrMEMs) is used to extract the feature from chest X-ray images. An orthogonal Exponent moments of fractional-orders is derived to extract 961 high accurate features from each COVID-19 input image. These moments represents digital images for low and high orders and are also robust to noise. Further extracted features are classified into training and testing datasets followed by a Manta-Ray Foraging Optimization and differential evolution (MRFODE) algorithm to remove the irrelevant features, and this is achieved by providing a set of solutions and computing the fitness value for each of them. The k-nearest neighbors (KNN) classifier is then used to select the best of them based on training sets, i.e., distinguish between normal and COVID-19 people, and the workflow is shown in Figure 1.5.

In this study, two different datasets were collected: the first dataset consists of 216 COVID-19 positive and 1,675 COVID-19 negative images (Cohen, Morrison, and Dao 2020), and the second dataset is composed of 219 COVID-19 positive images and 1,341 COVID-19 negative images (Chowdhury et al. 2020). They were then compared with the proposed method with the other deep neural network (DNN), and it was found that the proposed method produces better results in the classification criteria with just 16–18 features and with an accuracy of 96.09% and 98.09% for the first and second datasets, respectively, than the DNN with a 50 K feature set.

In another study (Apostolopoulos and Mpesiana 2020), a procedure called "transfer learning" is employed with a convolutional neural network for the automatic detection of COVID-19 from a set of thoracic X-ray images. So they have collected data of peoples suffering from bacterial pneumonia, confirmed cases of COVID, and some healthy people and further utilized it for the automatic COVID-19 detection and evaluated the performance of the state-of-the-art CNN for the classification of medical images. A total of 1,427 thoracic X-ray images were collected. In these collected data, 700 images were of bacterial pneumonia, 224 images of confirmed COVID cases, and 504 X-ray images of healthy patients were used to train and test the CNN model. Among all datasets, the sample size of COVID-19 is very limited, and, as a result, there is a need for transfer learning to train deep CNNs. Transfer learning is a technique where prior knowledge mined in CNN from a given data is transferred to solve the related task differently with novel datasets, and they are usually of small sample size (Weiss, Khoshgoftaar, and Wang 2016). Due to this fact, the state-of-the-art method is citified and requires extensive data. CNN is used for

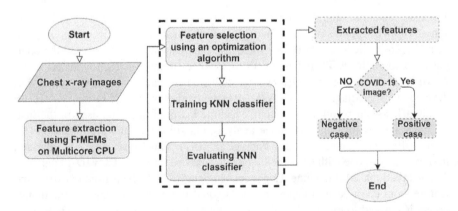

FIGURE 1.5 Flowchart to classify COVID-19 images using KNN classifier. (Elaziz et al. 2020.)

data classification, and the transfer learning parameter is a layer cutoff in which the amount of untrained data is placed from the bottom of the CNN, and those layers that are similar to the output are made workable, and other neural networks are used as a classifier and positioned at the top of the CNN. From the experimental results, it is concluded that transfer learning with X-ray images provides required biomarkers related to COVID-19 disease with 96.78% accuracy, 98.66% sensitivity, and 96.46% specificity.

1.2.4 COVID Patient's Health Prediction Using Machine Learning

One of the essential characteristics of COVID-19 is that it spreads via human-to-human interaction. It also spreads from patients to healthcare staff due to contact transmission, and so wireless infrastructure, real-time collection of data, and processing to the end-user are highly required these days. In a study by Iwendi et al. (2020), different ML algorithms are used, replacing the traditional healthcare system to process healthcare and travel data of COVID-positive patients in Wuhan, to predict COVID patients based on symptoms, travel history, and lag in reporting the case from the previous data of the patient. The various classifiers used to train the model are Decision Tree Classifier, Support Vector Classifier, Gaussian Naïve Bayes Classifier, and Boosted Random Forest Classifier. The model's performance is evaluated based on accuracy, precision, recall, and F1 ranking (Apostolopoulos and Mpesiana 2020).

Correlation between the features of the datasets founded that the Boosted Random Forest algorithm is the best performing model boosted by AdaBoost algorithm, and the F1 Score is 0.86 on the COVID-19 patient dataset, and so it was concluded that this algorithm provides correct predictions even with unevenly distributed datasets and found that the death rate is high in Wuhan natives than non-natives of Wuhan.

1.2.5 Machine Learning in Survival Analysis and Discharge Time Prediction of COVID-19 Patients

The outbreak of this pandemic has pressurized the world's healthcare systems, due to which the demand for medical care requirements has tremendously increased. A limited number of beds are available in hospitals, and intensive care units (ICUs) are also restricted. In such a situation, a quick estimation of the period of stay of COVID patients will reduce the workload and help in the proper management of healthcare availability. One of the study (Sonja A. Rasmussen and Smulian 2020) employs a statistical model and machine learning approach in predicting how long a patient will stay in hospital and evaluate the patient's clinical impact. Here, several computational analyses have been done on 1,182 patients based on age and gender to analyze their survival characteristics. The dataset consists of a censored dataset, so model performance in predicting discharge time is compared with the c-index (concordance index). C-index assesses the prediction of algorithms used for survival analysis by calculating the percentage of concordant pairs among the evaluating pairs (Uno et al. 2011). Here seven different algorithms are used for the evaluation, namely IPCRidge, CoxPH, Coxnet, Stagewise Gradient Boosting (GB), Component-wise GB, Fast

SVM, and Fast Kernel SVM. Among these, the IPCRidge has the least accuracy. As per the result, a linear function is relevant in the feature and survival time analysis. Therefore, considering a boosting method in which Stagewise GB process beats the other algorithms in discharge time predictions in terms of accuracy, age and gender of patient and model features. According to the Kaplan–Meier (KM) and Cox regression method, age and gender of hospitalized patients directly affect their recovery time, and male gender and elderly peoples are associated with low hospital discharge probabilities. This study provides a baseline for recovery rate prediction, which will help future research and the healthcare system.

1.2.6 Machine Learning in Contact Tracing of COVID-19

The most crucial step for controlling coronavirus's spread is contact tracing, as the virus spreads from one person to another through saliva and droplets (www.who.int). Contact tracing plays a vital role in the healthcare system. It will help identify and manage the people with COVID-19 and can suppress the outbreak of a pandemic throughout a population (Lalmuanawma, Hussain, and Chhakchhuak 2020). Nowadays, digital contact tracing methods are used by many countries, notably South Korea and Singapore (Wong, Leo, and Tan 2020), such as mobile data tracing, GPS (Global Positioning System), proximity tool Bluetooth, etc. These methods allow quicker processing of data than non-digital systems. The digital tracing process employs machine learning and artificial tools for the analysis of the disease. Several countries have employed ML and AI in digital tracing for infectious chronic wasting disease (Rorres et al. 2018) using centralized, decentralized, or hybrid techniques to minimize traditional, labor-intensive, and manual tracing methods. One study (Ferretti et al. 2020) highlighted the challenges and voluntariness of COVID-19 tracking apps (CTAs) that provide information about testing or advice for self-isolation from healthcare experts. A schematic diagram for COVID-19 contact tracing based on apps is shown in Figure 1.6.

The challenges for CTAs are related to the effectiveness and correct use of the apps, as they require a minimum of 60% of the population to participate (Ferretti et al. 2020). Also, with technical problems, privacy issues like detecting false-positive cases, and security risks and inequities like the unequal distribution of CTAs among the population, are detected (Klar and Lanzerath 2020; Leprince-Ringuet 2020; Luciano 2020). Each step of digital contact tracing using CTAs must be voluntary, including the decision to carry a smartphone, installing the app on a smartphone, leaving the CTA operating in the background all the time, responding to its alerts, and sharing contact of a COVID positive case (Dubov and Shoptawb 2020). Hence, this experiment's findings can be utilized and reviewed periodically and shared openly for design improvement and proper implementation in the future.

1.3 CONCLUSION

The outbreak of COVID-19 has harassed the lives of millions of people at both the social and personal levels, as many people have lost their jobs and their lives. It

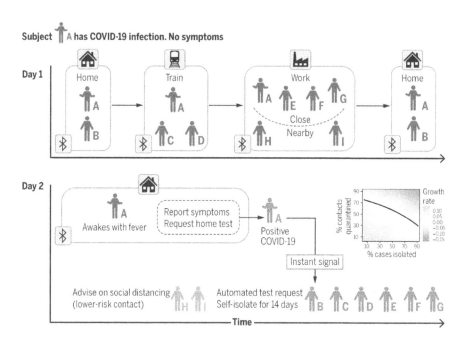

FIGURE 1.6 Schematic of app-based COVID-19 contact tracing. (Ferretti et al. 2020.)

has also triggered the world's biggest economic shock in decades, as most countries have forced full restrictions on movement to avoid or reduce the virus's transmission. These crises are demanding urgent action to bolster healthcare and economic systems and protect the world's vulnerable population. Researchers from various countries are providing all their efforts to forecast, predict, and develop medicine/vaccines to fight against these challenging situations. This comprehensive survey has addressed researchers' recent machine learning techniques against the COVID-19 pandemic. First, we have provided a basic introduction to the COVID-19 disease, about its structure and spreading mechanisms. The chapter has thus far addressed machine learning motivation for rapid approaches that effectively fight against disease, presented various applications of machine learning for the prediction, detection, diagnosis, forecasting, and contact tracing of the outbreak, and discussed some of the challenges that are to be faced and improved on in machine learning. Finally, it is found that machine learning has significantly improved treatment and medication processes and has lessened human interventions in healthcare systems, and is still up to the mark in tackling the COVID-19 pandemic.

REFERENCES

Apostolopoulos, Ioannis D., and Tzani A. Mpesiana. 2020. "Covid-19: Automatic Detection from X-Ray Images Utilizing Transfer Learning with Convolutional Neural Networks." *Physical and Engineering Sciences in Medicine* 43 (2). Springer International Publishing: 635–40. doi:10.1007/s13246-020-00865-4.

Asri, Hiba, Hajar Mousannif, Hassan Al Moatassime, and Thomas Noel. 2016. "Using Machine Learning Algorithms for Breast Cancer Risk Prediction and Diagnosis." *Procedia Computer Science* 83 (Fams). Elsevier Masson SAS: 1064–69. doi:10.1016/j.procs.2016.04.224.

Chakraborty, Tanujit, and Indrajit Ghosh. 2020. "Real-Time Forecasts and Risk Assessment of Novel Coronavirus (COVID-19) Cases: A Data-Driven Analysis." *Chaos, Solitons, and Fractals* 135. Elsevier Ltd. doi:10.1016/j.chaos.2020.109850.

Chaurasia, Vikas, and Saurabh Pal. 2020. "COVID-19 Pandemic: Application of Machine Learning Time Series Analysis for Prediction of Human Future." *SSRN Electronic Journal*, 1–16. doi:10.2139/ssrn.3652378.

Chouhan, Vikash, Sanjay Kumar Singh, Aditya Khamparia, Deepak Gupta, Prayag Tiwari, Catarina Moreira, Robertas Damaševičius, and Victor Hugo C. de Albuquerque. 2020. "A Novel Transfer Learning Based Approach for Pneumonia Detection in Chest X-Ray Images." *Applied Sciences (Switzerland)* 10 (2). doi:10.3390/app100 20559.

Chowdhury, Muhammad E.H., Tawsifur Rahman, Amith Khandakar, Rashid Mazhar, Muhammad Abdul Kadir, Zaid Bin Mahbub, Khandakar Reajul Islam, et al. 2020. "Can AI Help in Screening Viral and COVID-19 Pneumonia?" *IEEE Access* 8: 132665–76. doi:10.1109/ACCESS.2020.3010287.

Cohen, Joseph Paul, Paul Morrison, and Lan Dao. 2020. "COVID-19 Image Data Collection." http://arxiv.org/abs/2003.11597.

Dubov, Alex, and Steven Shoptawb. 2020. "The Value and Ethics of Using Technology to Contain the COVID-19 Epidemic." *American Journal of Bioethics* 20 (7): W7–11. doi:10.1080/15265161.2020.1764136.

Elaziz, Mohamed Abd, Mohamed Abd Elaziz, Khalid M. Hosny, Ahmad Salah, Mohamed M. Darwish, Songfeng Lu, and Ahmed T. Sahlol. 2020. "New Machine Learning Method for Imagebased Diagnosis of COVID-19." *PLoS ONE* 15 (6 June). doi:10.1371/journal.pone.0235187.

Fan, Yi, Kai Zhao, Zheng Li Shi, and Peng Zhou. 2019. "Bat Coronaviruses in China." *Viruses* 11 (3): 27–32. doi:10.3390/v11030210.

Ferretti, Luca, Chris Wymant, Michelle Kendall, Lele Zhao, Anel Nurtay, Lucie Abeler-Dörner, Michael Parker, David Bonsall, and Christophe Fraser. 2020. "Quantifying SARS-CoV-2 Transmission Suggests Epidemic Control with Digital Contact Tracing." *Science* 368 (6491): 0–8. doi:10.1126/science.abb6936.

Iwendi, Celestine, Ali Kashif Bashir, Atharva Peshkar, Jyotir Moy Chatterjee, Swetha Pasupuleti, Rishita Mishra, Sofia Pillai, and Ohyun Jo. 2020. "COVID-19 Patient Health Prediction Using Boosted Random Forest Algorithm." *Frontiers in Public Health* 8 (July): 1–9. doi:10.3389/fpubh.2020.00357.

Jamshidi, Mohammad, Ali Lalbakhsh, Jakub Talla, Zdenek Peroutka, Farimah Hadjilooei, Pedram Lalbakhsh, Morteza Jamshidi, et al. 2020. "Artificial Intelligence and COVID-19: Deep Learning Approaches for Diagnosis and Treatment." *IEEE Access* 8 (December 2019): 109581–95. doi:10.1109/ACCESS.2020.3001973.

Ke, Qiao, Jiangshe Zhang, Wei Wei, Dawid Połap, Marcin Woźniak, Leon Kośmider, and Robertas Damaševčius. 2019. "A Neuro-Heuristic Approach for Recognition of Lung Diseases from X-Ray Images." *Expert Systems with Applications* 126: 218–32. doi:10.1016/j.eswa.2019.01.060.

Klar, Renate, and Dirk Lanzerath. 2020. "The Ethics of COVID-19 Tracking Apps – Challenges and Voluntariness." *Research Ethics*. doi:10.1177/1747016120943622.

Lalmuanawma, Samuel, Jamal Hussain, and Lalrinfela Chhakchhuak. 2020. "Applications of Machine Learning and Artificial Intelligence for Covid-19 (SARS-CoV-2) Pandemic: A Review." *Chaos, Solitons, and Fractals* 139. Elsevier Ltd: 110059. doi:10.1016/j.chaos.2020.110059.

Lapuerta, Pablo, Stanley P. Azen, and Laurie LaBree. 1995. "Use of Neural Networks in Predicting the Risk of Coronary Artery Disease." *Computers and Biomedical Research.* doi:10.1006/cbmr.1995.1004.

Leprince-Ringuet, D. 2020. "Contact-Tracing Apps: Why the NHS Said No to Apple and Google's Plan. ZDNet." https://www.zdnet.com/article/contact-tracingapps-why-the-nhs-said-no-to-apple-and-googles-plan/.

Luciano, Floridi. 2020. "Mind the App—Considerations on the Ethical Risks of COVID-19 Apps." *Philosophy and Technology.* doi:10.1007/s13347-020-00408-5.

Nishiga, Masataka, Dao Wen Wang, Yaling Han, David B. Lewis, and Joseph C. Wu. 2020. "COVID-19 and Cardiovascular Disease: From Basic Mechanisms to Clinical Perspectives." *Nature Reviews. Cardiology* 17 (9). Springer US: 543–58. doi:10.1038/s41569-020-0413-9.

Rampal, Lekhraj, and Liew Boon Seng. 2020. "Coronavirus Disease (COVID-19) Pandemic." *Medical Journal of Malaysia.* https://www.who.int/emergencies/diseases/novel-coronavirus-2019/events-as-theyhappen.

Rasmussen, Sonja A., and John C. Smulian. 2020. "Since January 2020 Elsevier Has Created a COVID-19 Resource Centre with Free Information in English and Mandarin on the Novel Coronavirus COVID-." *Annals of Oncology,* January: 19–21. doi:10.1007/s00134-020-05991-x.Bizzarro.

Rorres, Chris, Maria Romano, Jennifer A. Miller, Jana M. Mossey, Tony H. Grubesic, David E. Zellner, and Gary Smith. 2018. "Contact Tracing for the Control of Infectious Disease Epidemics: Chronic Wasting Disease in Deer Farms." *Epidemics* 23 (December 2017). Elsevier: 71–75. doi:10.1016/j.epidem.2017.12.006.

Rustam, Furqan, Aijaz Ahmad Reshi, Arif Mehmood, Saleem Ullah, Byung Won On, Waqar Aslam, and Gyu Sang Choi. 2020. "COVID-19 Future Forecasting Using Supervised Machine Learning Models." *IEEE Access* 8: 101489–99. doi:10.1109/ACCESS.2020.2997311.

Uno, Hajime, Tianxi Cai, Michael J. Pencina, Ralph B. D'Agostino, and L. J. Wei. 2011. "On the C-Statistics for Evaluating Overall Adequacy of Risk Prediction Procedures with Censored Survival Data." *Statistics in Medicine* 30 (10): 1105–17. doi:10.1002/sim.4154.

Vaishya, Raju, Mohd Javaid, Ibrahim Haleem, and Abid Haleem. 2020. *Since January 2020 Elsevier Has Created a COVID-19 Resource Centre with Free Information in English and Mandarin on the Novel Coronavirus COVID-19. The COVID-19 Resource Centre Is Hosted on Elsevier Connect, the Company's Public News and Information no.* January.

Weiss, Karl, Taghi M. Khoshgoftaar, and Ding Ding Wang. 2016. "A Survey of Transfer Learning." *Journal of Big Data* 3. Springer International Publishing. doi:10.1186/s40537-016-0043-6.

WHO. 2020. "WHO Coronavirus Disease (COVID-19) Dashboard | WHO Coronavirus Disease (COVID-19) Dashboard." *Who.* https://covid19.who.int/

Wong, John E.L., Yee Sin Leo, and Chorh Chuan Tan. 2020. "COVID-19 in Singapore - Current Experience: Critical Global Issues That Require Attention and Action." *JAMA: The Journal of the American Medical Association* 323 (13): 1243–44. doi:10.1001/jama.2020.2467.

Yang, Lin, Yizhe Zhang, Jianxu Chen, Siyuan Zhang, and Danny Z. Chen. 2017. "Suggestive Annotation: A Deep Active Learning Framework for Biomedical Image Segmentation." *Lecture Notes in Computer Science (Including Subseries Lecture Notes in Artificial Intelligence and Lecture Notes in Bioinformatics)* 10435 LNCS (1): 399–407. doi:10.1007/978-3-319-66179-7_46.

Zaki, Ali Moh, Sander Van Boheemen, Theo M. Bestebroer, Albert D.M.E. Osterhaus, and Ron A.M. Fouchier. 2012. "Isolation of a Novel Coronavirus from a Man with Pneumonia in Saudi Arabia." *New England Journal of Medicine* 367 (19): 1814–20. doi:10.1056/NEJMoa1211721.

2 Analysis of COVID-19 Data Using Consensus Clustering Technique

Arko Banerjee, Sunandana Mukherjee, Chhabi Rani Panigrahi, Bibudhendu Pati, and Rajib Mall

CONTENTS

2.1	Introduction	17
2.2	Literature Study	19
2.3	Methodology Used	19
	2.3.1 Clustering	19
	2.3.2 Consensus Clustering	20
	2.3.3 Problem Formulation	20
	2.3.4 Pairwise-Similarity-Based Consensus Clustering Method	21
2.4	Proposed Method	22
2.5	Experimental Setup and Result Analysis	23
2.6	Conclusion	26
References		26

2.1 INTRODUCTION

Viruses are intracellular parasites whose life cycle depends on hijacking cellular functions to help spread their replication. Coronaviruses are a large family of viruses that have a wide range of hosts including humans (Paules et al., 2020). The external structure that features little crown-like spikes gives the virus family its name. Usually human coronaviruses are benign and can cause mild respiratory illnesses like the common cold. Coronaviruses that are able to transmit from an animal to a human (this process is called zoonosis or zoonotic transmission) (Zoonoses, 2020) via mutation can cause serious illness due to lack of immunity to the new viruses in humans. The three well-known coronaviruses that are able to cause zoonosis are: Severe Acute Respiratory Syndrome Coronavirus (SARS-CoV) (SARS, 2020), Middle East Respiratory Syndrome Coronavirus (MERS-CoV) (MERS-CoV, 2020), and SARS-CoV-2, the virus that caused Coronavirus Disease 2019 (COVID-19) (WHO, 2020).

In order to infect and replicate, all viruses must bind themselves to specific receptor molecules on the surface of target cells. A recent study (Lu et al., 2020) has confirmed that both SARS-CoV-2 and SARS-CoV use the same host cell receptor. Due to this similarity, the International Committee on Taxonomy of Viruses (ICTV), on 11 February 2020, stated the name of the new virus as "Severe Acute Respiratory Syndrome Coronavirus 2 (SARS-CoV-2)" (ICTV, 2020). Though these two viruses are related, they differ significantly in the processes of transmission and receptor binding. In the case of SARS-CoV-2, the amount of viral load appears to be present in the highest amount in the nose and throat of infected people shortly after symptoms develop (Zou et al., 2020). But, in the case of SARS-CoV, the viral loads accumulate significantly in the later stage of infection. This indicates that people infected with COVID-19 start spreading the virus while the symptoms are barely noticeable or developing. This is the reason why it is more difficult to stop SARS-CoV-2 from spreading than SARS-CoV. A recent article (Tai et al., 2020) states that SARS-CoV-2 binds to the host cell receptor with a higher affinity than SARS-CoV, which could be also a reason for its higher rate of infection than the SARS virus.

As per WHO's report on 26 March 2020, the zoonotic source of SARS-CoV-2 was unknown, but the first human cases of COVID-19 were detected in Wuhan City in China back in December 2019 (Situation Report – 94, 2020). At present, nearly all countries have been affected by COVID-19. As of 10 September 2020 (02:00 IST), 27.97 million people were reported to have been affected, 19.98 million (71.45%) recovered, and 9,06,211 (3.24%) died (Worldometer, 2020; MoHFW, 2020; India Situation Reports, 2020). So far there is no specific treatment or vaccine that has been reported to cure COVID-19. However, there are several ongoing clinical trials in both Western and traditional medicines. In India, on the basis of patients' clinical condition, allopathic, homeopathic, or Ayurvedic remedies are applied to provide comfort and alleviate symptoms of mild to moderate COVID-19 infection.

In the absence of definite treatment or vaccine, machine learning (ML) and related technologies have been playing a crucial role in designing predictive and preventive models to contain pandemics like COVID-19. From modeling the pandemic and preventing the spread of the disease to diagnosis and treatment, academic institutions and technology companies in India have been actively involved in developing solutions by leveraging computational intelligence to support the government in this fight against COVID-19.

In this study, we have utilized a data mining technique called *Consensus Clustering* or *Clustering Ensemble* to analyze the COVID-19 data of India. The key objective of this chapter is to figure out the relative performances of all the states and union territories (UTs) in India hit by COVID-19 in handling the pandemic when the lockdown has been relaxed on a stage-by-stage basis. The analysis presented in this chapter will be useful for the state governments and corona warriors (doctors, police, and others involved) to optimize government policies, decisions, medical facilities, treatment, etc., to reduce the spread of COVID-19 cases.

The rest of this chapter is structured as follows. Section 2.2 reviews the literature. Section 2.3 presents the methodology used in this chapter and Section 2.4 presents our proposed method. Empirical outcomes on COVID-19 data of India are reported in Section 2.5, and Section 2.6 concludes the chapter.

2.2 LITERATURE STUDY

We give a brief overview of the related work reported in the literature on data analysis of COVID-19 using ML techniques. Sujath et al. (2020) presented a model using ML algorithms such as linear regression, multilayer perceptron, and vector autoregression on Kaggle COVID-19 data of India to predict the spread of the pandemic in the not-so-distant future. The study concluded that the multilayer perceptron method produced better prediction results than the linear regression and vector autoregression methods. Benvenuto et al. (2020) proposed an Auto-Regressive Integrated Moving Average (ARIMA) model on the Johns Hopkins epidemiological data to predict the spread of COVID-19. The paper concluded that there was enough evidence that the virus had developed new mutations, otherwise the graph of the number of confirmed cases should have reached a plateau.

Deb et al. (2020) proposed a time series model to examine the pattern of the spread of COVID-19 in different countries and revealed that the spread of the pandemic was strongly captured by a time-dependent quadratic trend. Rajan et al. (2020) used ML models such as the SEIR model and regression model to predict the spread of COVID-19 in India during the time period of 30 January to 10 May 2020.

Zlatan et al. (2020) trained multilayer perceptron (MLP) and artificial neural network (ANN) with a publicly available time series COVID-19 worldwide dataset (22 January–12 March 2020) to predict the maximal number of patients at each time unit. Iwendi et al. (2020) proposed a Random Forest model that predicted the scale of COVID-19 infections on the basis of patients' health and demographic data. The model scored an accuracy of 94% on the considered dataset. Ardabili et al. (2020) used two models on multi-layered perceptron and adaptive network-based fuzzy inference system to measure variation in the behavior of the COVID-19 outbreak from nation to nation.

2.3 METHODOLOGY USED

In this section, we describe the methodology called "clustering analysis" that is used in this chapter to analyze COVID-19 data.

2.3.1 CLUSTERING

Clustering is an unsupervised ML technique that concerns the grouping of data points. Given a collection of data points, we can use a clustering algorithm to partition the data points into mutually exclusive and exhaustive sets in such a way that data points in one set have similar properties and/or attributes, while they are dissimilar with data points in other sets. Clustering is a common technique in data science to understand the pattern of distribution of data and hence to gain valuable insights from the data. Let us consider a real-life application of clustering. Suppose the head of a rental store wants to understand the pattern of preferences of his customers so that they can provide better facility to the customers. But, since there can be thousands of customers for a big retail store, it is not feasible to study the behavior of each customer; instead, the owner can make a clustering of the entire customer datasets on the basis of the attributes representing their preferences. As a result, in each cluster in the clustering, the customers should have a similar kind of buying behavior compared to customers belonging to different clusters. Now the owner can select a representative customer from each cluster who should be the customer that is most similar to other members of the cluster in which that representative belongs. Hence the owner can understand the difference of preferences of a group of customers by considering only the representative customer from each cluster and devise a unique business strategy for each one of them. Like this, clustering has a large number of applications in real life. Some of the popular applications of clustering in different domains include: recommender systems, social network analysis, identifying objects in medical imaging, image segmentation, anomaly detection, etc. Since clustering is done on the basis of similarity of data points, which is subjective in nature, as a result there exist different types of concepts or approaches in performing clustering of data in the literature (Britannica, 2020).

In this chapter, we implement an optimization technique called *Consensus Clustering or Clustering Ensemble* which utilizes two popular clustering algorithms – K-means clustering (MacQueen, 1967) and Agglomerative Hierarchical clustering (Ward, 1963), to analyze the COVID-19 situation in India.

2.3.2 Consensus Clustering

The necessity of consensus clustering emerges because there does not exist a universal clustering algorithm that can provide acceptable results for any dataset (Kleinberg, 2002). Some clustering algorithms also give different clustering results based on different values of initial parameters, and there is no way to determine the most appropriate values of these parameters for a given situation (Jain and Dubes, 1988). In such situations, consensus clustering attempts to combine the results of different input clusterings (referred to as base clusterings), obtained by running different clustering algorithms or by running a single clustering algorithm with different parameter initialization, in order to get a qualitatively better and robust clustering solution (WHO, 2020). In our analysis, we generate the ensemble (collection) of base clusterings using the K-means algorithm (MacQueen, 1967) with different parameter initialization. In the next subsection we present the formulation of the consensus clustering problem.

2.3.3 Problem Formulation

Let D be a dataset for clustering that contains N number of data points, $D = \{d_1, d_2, \ldots, d_N\}$, where each data point is an m-dimensional vector of real numbers, i.e. $d_t \in R^m$, $t = 1, 2, \ldots, N$. Let E be an ensemble of M number of base clusterings (also called partitions) of D. We write $E = \{P_1, P_2, \ldots, P_M\}$, where each clustering P_t ($t = 1, 2, \ldots, M$) is the set of K number of clusters that divides the data points of the entire dataset D into K number of mutually exclusive and exhaustive sets. To find the optimal number of clusters K, there are different methods proposed in the literature, such as Elbow (Thorndike, 1953), Silhouette coefficient (Rousseeuw, 1987) and Bayesian Information Criterion (BIC) (Schwarz, 1978). In this work, we used the popular Elbow method to predetermine the number of clusters for the K-means algorithm.

We write $P_t = \{C_1^t; C_2^t, \ldots, C_K^t\}$, where C_i^t is the ith cluster with $C_i^t \subseteq D$, $C_i^t \cap C_j^t = \emptyset$, ($i \neq j$) and $\cup_{i=1}^{K} C_i^t = D$ ($i, j = 1, 2, \ldots, K$). The aim of consensus clustering is to discover a point in the sample space of clusterings which is the median of all the given base clusterings in E. According to different scientific requirements, various approaches of consensus clustering methods are proposed in the literature (Vega-Pons and Ruiz-Shulcloper, 2011). In this chapter, we used the pairwise-similarity-based consensus clustering method for implementation.

2.3.4 Pairwise-Similarity-Based Consensus Clustering Method

In pairwise-similarity-based method, each base clustering is mapped into a co-association matrix depicting similarity between all pairs of input data points regarding their belongingness in different clusters in the ensemble. Finally, a clustering algorithm is applied to the co-association matrix to obtain the consensus clustering. Fred et al. (2002) agglomerated clusterings, due to multiple runs of K-means algorithm, into a co-association matrix, and the final consensus clustering was derived from it by applying a hierarchical single-link algorithm (Ward, 1963).

Let $C_.^t$ be the cluster in which data instance d_i in D belongs in clustering P_t; we compute the co-association matrix $A = \{a_{ij}\}_{N \times N}$ as follows:

$$aij = \frac{1}{M} \cdot \sum_{\substack{t=1 \\ d_i \in C_.^t}}^{M} W(C_.^t) \cdot \delta_{ij}^t \text{ where, } \delta_{ij}^t = \begin{cases} 1, & \text{if } d_j \in C_.^t \\ 0, & \text{otherwise} \end{cases}$$

Here "." is used as a subscript in the cluster notation that means the label of the cluster in the clustering is unimportant for the said definition. In Figure 2.1, a co-association matrix (A) is generated from two clusterings of a dataset that contains four data points. Finally the consensus clustering is derived from the co-association matrix using Hierarchical Agglomerative Clustering.

Hierarchical Agglomerative Clustering (HAC) is a well-known clustering technique that takes a co-association matrix as input, and by initializing each data instance as a cluster, it iteratively performs pairwise concatenation (agglomeration) of most similar clusters to achieve a hierarchical representation of the final clustering. Out of certain approaches in HAC which are used to calculate the similarity between two clusters, we consider the group average (also referred to as average linkage) similarity approach. We choose the group average approach as it is efficient in separating clusters if there is noise between them. In this approach, the

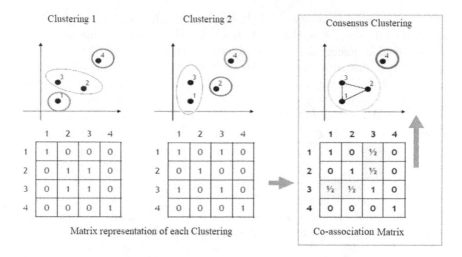

FIGURE 2.1 A toy example of co-association matrix.

similarity between two clusters is measured by taking the average of the similarities between every pair of data points belonging to two clusters, respectively. In our implementation, the similarity between data points is stored in the co-association matrix. Mathematically, the group average similarity between two clusters C_s and C_t can be written as:

$$\text{sim}(C_s, C_t) = \frac{1}{|C_s| \cdot |C_t|} \sum_{d_i \in C_s, d_j \in C_t} a_{ij}$$

Without loss of generality, let us assume that at a certain iteration, clusters C_s and C_t have the highest similarity among all other existing pairs of clusters. Then they are concatenated into a bigger new cluster $C_s \cup C_t$, reducing the total number of clusters by one. In the next iteration, the similarity of the new cluster with any cluster, say C_u, $\text{sim}(C_s \cup C_t, C_u)$, can be computed as follows:

$$\text{sim}(C_s \cup C_t, C_u) = \frac{1}{(|C_s \cup C_t|) \cdot |C_u|} \sum_{d_i \in C_s \cup C_t, d_j \in C_u} a_{ij}$$

The said concatenation process of clusters is repeated until the number of clusters comes down to the optimal number of clusters K.

2.4 PROPOSED METHOD

In India, out of 36 states and union territories (UTs), 35 had been affected by COVID-19 on 1 July 2020. The only UT in which COVID-19 had not been detected was Lakshadweep. With the help of the Central Government of India, the government of each state and UT had been taking the necessary steps to contain the pandemic.

In our study, we make a relative comparison of performances of the states and UTs during the month of June 2020, which can be further extended to any period of time without changing the method used. We select the month of June 2020 since during this month the unlock process in India started after more than two months of lockdown. A day-to-day state-wise dataset of the COVID-19 situation is collected from the portal of the Ministry of Health of the Government of India (MoHFW, 2020) that contained three attributes, namely the number of infected, recovered, and deceased persons due to COVID-19. To define the performance index of each state or UT, we update the second and third attributes as the percentage of non-recovered and percentage of deceased persons with respect to the infected ones (first attribute), respectively. We normalize the three attributes of all states and UTs by subtracting the corresponding means and dividing by the corresponding standard deviations. A performance index of a state or UT can then be measured as the sum of all its three attributes. A relatively increasing performance index value indicates decreasing performance of the state or UT in combating COVID-19 on a specific day. On the basis of these three normalized attributes, we perform the consensus clustering technique, which we discussed in Section 2.3, on the said day-to-day state-wise data of COVID-19. As a result, we achieve a robust clustering, containing clusters of states and UTs as members. Members of a cluster are the similar performing states and UTs with regard to the three attributes of the day-wise COVID-19 data. The predefined number of clusters in the clustering is determined by the Elbow method and in the empirical section we suggested seven optimum numbers of clusters. We define the performance of each cluster as the average of the performance index values of its members. All the clusters in the clustering are sorted with respect to their performance values and relabeled accordingly. If there are seven clusters in the clustering, a clustering having a label of 1 would be the best performing cluster of states and UTs, whereas the cluster labeled 7 would be the worst-performing groups of states or UTs.

The said cluster performance evaluation process is repeated for each day-wise data during the entire month of June 2020. On the basis of this, the monthly performance of the states and UTs are obtained. The monthly performance of the states and UTs is measured with three parameters – the mean of the cluster labels, Shannon's entropy (Shannon, 1948) of the cluster labels, and improvement (IM) in cluster labels, which is the difference between the starting and ending cluster labels of the month. The lesser the mean value, the better the average monthly performance of the corresponding state or UT. The lesser the entropy value, the more robust or stable the monthly performance. The more the value of IM, the more the corresponding state or UT improves in its relative position among other states, at the end of the month with respect to its position at the beginning. The proposed method is shown stepwise in Figure 2.2.

2.5 EXPERIMENTAL SETUP AND RESULT ANALYSIS

The experiment is repeated twice on two different representations of the first attribute of the dataset, which is the number of infected persons. In the first trial of the experiment, the data is taken as defined in Section 2.4. In the second trial, population

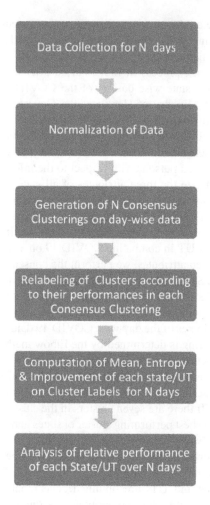

FIGURE 2.2 Steps of the proposed method.

density per square kilometer is considered for each state/UT (NITI, 2020). If two states have the same number of infected persons, then the state having greater population density can be considered to have performed better in restricting the spread of COVID-19. In this case, we update the first attribute in the data with the unnormalized number of infected persons divided by the population density per square kilometer of the corresponding state/UT. Therefore, we perform two trials of the experiment – one "without considering population density" (first trial) and the other "with considering population density" (second trial). The Elbow method suggests $K = 7$ is the optimum numbers of clusters and K-means is run 10 times to generate an ensemble of base clusterings of size $M = 10$ for the consensus clustering on each day-wise dataset.

Table 2.1 presents the performances of each state/UT. With respect to mean value, the state of Maharashtra remains the worst hit state by COVID-19, whereas Andaman

TABLE 2.1
Experimental Outcome on COVID-19 Data of India during the Month of June 2020

State/UT of India	Without Considering Population Density			With Considering Population Density		
	Mean	Entropy	IM	Mean	Entropy	IM
Andaman and Nicobar Islands	1.103448	0.431017	−1	1.241379	0.869996	−1
Andhra Pradesh	2.482759	1.778016	−2	2.758621	1.930992	−2
Arunachal Pradesh	3.551724	0.992267	−1	3.689655	1.269921	1
Assam	2.034483	1.855312	2	2.206897	2.042002	3
Bihar	1.586207	1.361016	1	1.689655	1.372553	1
Chandigarh	1.413793	1.124259	1	1.448276	1.124259	1
Chhattisgarh	1.931034	1.807283	2	2.172414	1.997719	3
Dadra and Nagar Haveli and Daman and Diu	3.482759	1.178358	−1	3.689655	1.269921	1
Delhi	5.448276	0.992267	−1	4.689655	1.3732	1
Goa	3.241379	1.577799	−2	3.206897	1.736524	−1
Gujarat	5.551724	1.149684	1	5.551724	1.149684	2
Haryana	2.344828	1.436626	−1	2.413793	1.590571	−2
Himachal Pradesh	2	1.481901	0	2.137931	1.45907	0
Jammu and Kashmir	2.413793	1.553752	0	2.517241	1.69135	0
Jharkhand	1.655172	1.376384	1	1.724138	1.361016	1
Karnataka	2.241379	1.34163	0	2.517241	1.736524	−3
Kerala	2.172414	1.609257	0	2.241379	1.723063	0
Ladakh	2.793103	1.722653	0	4.586207	2.132617	−3
Madhya Pradesh	4.931034	1.855959	1	5.413793	1.391453	2
Maharashtra	7	0	0	7	0	0
Manipur	3.413793	1.292164	−1	3.551724	1.284777	1
Meghalaya	3.068966	1.986356	5	3.310345	1.986497	1
Mizoram	3.37931	1.284777	−3	3.448276	1.456012	−2
Nagaland	2.827586	1.868215	−1	2.931034	2.029047	1
Odisha	1.413793	1.178304	1	1.413793	1.124259	1
Puducherry	2.931034	1.884376	−2	3.517241	1.932993	−3
Punjab	1.689655	1.439868	−2	2.103448	1.733727	−3
Rajasthan	1.827586	1.611103	1	2.241379	1.982668	−1
Sikkim	3.241379	1.467549	1	3.517241	1.36113	2
Tamil Nadu	5.344828	1.149684	−1	2.689655	1.820744	−2
Telangana	4.241379	1.885119	0	4.517241	1.456012	0
Tripura	2.068966	1.759758	1	2.068966	1.876598	1
Uttar Pradesh	4.241379	1.986497	1	4.62069	1.285982	1
Uttarakhand	2.103448	1.863232	2	2.206897	1.982668	3
West Bengal	4.862069	1.869325	1	4.896552	1.664673	2

and Nicobar Islands remains the least hit UT in both the first and second trials of the experiments. The situation in Maharashtra is the worst as it has a maximum mean value with zero entropy, which implies that the state maintained the worst situation all through the month of June 2020. On the other hand, Andaman and Nicobar Islands is one of the most consistently performing places with a smaller entropy value. Due to having the highest entropy values, the most fluctuation in performance is observed in the case of Uttar Pradesh (in the first trial) and Ladakh (in the second trial). The least improvement is observed in the case of Mizoram (first trial) and Ladakh, Karnataka, Punjab, and Puducherry (second trial). The most improvement is observed in the case of Meghalaya (first trial) and Assam, Chhattisgarh, and Uttarakhand (second trial).

2.6 CONCLUSION

In this chapter, we have used the consensus clustering technique to classify Indian states and UTs on the basis of the status of COVID-19. We have designed an unsupervised robust method to measure the relative performance of all the affected states and UTs in handling COVID-19 during the month of June 2020, when lockdown had been relaxed on a stage-by-stage basis. The analysis may help understand the impact of unlock in each state and UT and in turn may help the respective governments to curb the further spread of COVID-19 in India.

REFERENCES

Ardabili, S., Mosavi, A., Ghamisi, P., Ferdinand, F., Varkonyi-Koczy, A., Reuter, U., Rabczuk, T., and Anson, P. 2020. COVID-19 Outbreak Prediction with *Machine Learning*. doi:10.20944/preprints202004.0311.v1.

Benvenuto, D., Giovanetti, M., Vassallo, L., Angeletti, S., and Ciccozzi, M. 2020. Application of the ARIMA Model on the COVID-2019 Epidemic Dataset. *Data in Brief*, 29.

Britannica. 2020. Coronavirus Group. https://www.britannica.com/science/coronavirus-virus-group. Accessed 10 Sept 2020.

Deb, S., and Majumdar, M. 2020. A Time Series Method to Analyze Incidence Pattern and Estimate Reproduction Number of COVID-19, arXiv preprint arXiv:2003.10655

Fred, A.L.N., and Jain, A.K. 2002. Data Clustering Using Evidence Accumulation. *Proc. of the 16th International Conference on Pattern Recognition*, pp. 276–280.

ICTV (International Committee on Taxonomy of Viruses). https://talk.ictvonline.org. Accessed 10 Sept 2020.

India Situation Reports, Coronavirus Disease (covid-19) – WHO. https://www.who.int/india/emergencies/coronavirus-disease-(covid-19). Accessed 10 Sept 2020.

Iwendi, C., Bashir, A.K., Peshkar, A., Sujatha, R., Chatterjee, J.M., Pasupuleti, S., Mishra, R., Pillai, S., and O. Jo. 2020. COVID-19 Patient Health Prediction Using Boosted Random Forest Algorithm. *Frontiers in Public Health*. 8:2296–2565.

Jain, A.K., and Dubes, R.C. 1988. *Algorithms for Clustering Data. Engle-wood Cliffs*. New Jersey: Prentice Hall.

Kleinberg, J. 2002. An Impossibility Theorem for Clustering, *Proc. of Adv. in Neural Information Processing System*.

Lu, R., Zhao, X., Li, J., et al. 2020. Genomic Characterisation and Epidemiology of 2019 Novel Coronavirus: Implications for Virus Origins and Receptor Binding. *Lancet*. 395(10224):565–574.

MacQueen, J.B. 1967. Some Methods for Classification and Analysis of Multivariate Observations. *Proceedings of 5th Berkeley Symposium on Mathematical Statistics and Probability.* University of California Press. pp. 281–297.

MERS-CoV (Middle East Respiratory Syndrome Coronavirus) - WHO. https://www.who.int/news-room/fact-sheets/detail/middle-east-respiratory-syndrome-coronavirus-(mers-cov). Accessed 10 Sept 2020.

MoHFW (Ministry of Health and Family Welfare Government of India). 2020. https://www.mohfw.gov.in. Accessed 10 Sept 2020 at 02:55 IST.

NITI Aayog. 2020. https://www.niti.gov.in/niti/content/population-density-sq-km. Accessed 10 Sept 2020.

Paules, C.I., Marston, H.D., and Fauci, A.S. 2020. Coronavirus Infections-More Than Just the Common Cold. *JAMA.* 323(8):707–708.

Rajan, G., Gaurav, P., Poonam, C., and Saibal, K.P. 2020. *Machine Learning Models for Government to Predict COVID-19 Outbreak*, Association for Computing Machinery, New York, NY, Vol-1(4).

Rousseeuw, P.J. 1987. Silhouettes: A Graphical Aid to the Interpretation and Validation of Cluster Analysis. *Computational and Applied Mathematics.* 20:53–65.

SARS (Severe Acute Respiratory Syndrome) - WHO. https://www.who.int/ith/diseases/sars/en. Accessed 10 Sept 2020.

Schwarz, G.E. 1978. Estimating the Dimension of a Model. *Annals of Statistics.* 6(2):461–464.

Shannon, C.E. 1948. A Mathematical Theory of Communication. *Bell System Technical Journal.* 27(3):379–423.

Situation Report – 94, Coronavirus Disease 2019 (COVID-19) – WHO. https://www.who.int/docs/default-source/coronaviruse/situation-reports/20200423-sitrep-94-covid-19.pdf. Accessed 10 Sept 2020.

Sujath, R., Chatterjee, J.M., and Hassanien, A.E. 2020. A Machine Learning Forecasting Model for COVID-19 Pandemic in India. *Stochastic Environmental Research and Risk Assessment* 34:959–972.

Tai, W., He, L., Zhang, X., et al. 2020. Characterization of the Receptor-Binding Domain (RBD) of 2019 Novel Coronavirus: Implication for Development of RBD Protein as a Viral Attachment Inhibitor and Vaccine. *Cellular & Molecular Immunology* 17(6):613–620.

Thorndike, R.L. 1953. Who Belongs in the Family? *Psychometrika.* 18(4):267–276.

Vega-Pons, S., and Ruiz-Shulcloper, J. 2011. A Survey of Clustering Ensemble Algorithms. *IJPRAI.* 25(3):337–372.

Ward, J.H. 1963. Hierarchical Grouping to Optimize an Objective Function. *Journal of the American Statistical Association.* 58(301):236–244.

WHO. 2020. Naming the Coronavirus Disease (COVID-19) and the Virus That Causes it. https://www.who.int/emergencies/diseases/novel-coronavirus-2019/technical-guidance/naming-the-coronavirus-disease-(covid-2019)-and-the-virus-that-causes-it. Accessed 10 Sept 2020.

Worldometer. 2020. Countries in the World by Population. https://www.worldometers.info/world-opulation/population-by-country Accessed 10 Sept 2020 at 02:50 IST.

Zlatan, C., Sandi, B.Š., Nikola, A., Ivan, L., and Vedran, M. 2020. Modeling the Spread of COVID-19 Infection Using a Multilayer Perceptron. *Computational and Mathematical Methods in Medicine* 2020:Article ID 5714714.

Zoonoses – WHO. https://www.who.int/topics/zoonoses/en. Accessed 10 Sept 2020.

Zou, L., Ruan, F., Huang, M., et al. 2020. SARS-CoV-2 Viral Load in Upper Respiratory Specimens of Infected Patients. *The New England Journal of Medicine* 382(12):1177–1179.

3 MoBMGAN
Modified GAN-Based Transfer Learning for Automatic Detection of COVID-19 Cases Using Chest X-ray Images

Rajashree Nayak, Bunil Ku. Balabantaray and Dipti Patra

CONTENTS

- 3.1 Introduction 29
- 3.2 Literature Survey 30
- 3.3 Proposed Method..33
  - 3.3.1 Data Collection and Preprocessing..34
  - 3.3.2 Generation of Expanded Dataset via MGAN 34
  - 3.3.3 Fine-tuning Transfer Learning Models for Classification..................37
- 3.4 Result and Analysis 39
  - 3.4.1 Accuracy vs Computational Complexity 43
- 3.5 Conclusion 44
- References...44

3.1 INTRODUCTION

The novel coronavirus 2019 (COVID-19) disease caused by Severe Acute Respiratory Syndrome Coronavirus 2 (SARS-CoV-2) has been acknowledged as a wide-reaching pandemic by WHO. This pandemic has buckled the health-care systems of the whole world and has also drastically lowered the world economy. The exponentially increasing mortality rate of this dreadful disease across the world demands its early-stage detection. Early diagnosis of COVID-19 disease enables us to efficiently plan treatments and to take proper diagnostic decisions. This will preclude the rapid spread of this epidemic. Popularly used RT-PCR tool kits to detect the presence of the COVID-19 virus are sophisticated to handle, are costly, and have lower sensitivity rates which introduce false-negative results. Nevertheless, the detection process is further affected due to the dearth of expert physicians in remote areas. To resolve these issues, radio imaging (RI) techniques (X-ray and CT scan) are integrated with advanced artificial intelligence (AAI) techniques (deep learning models (DLMs) or transfer learning models (TLMs)) to provide superior performance in detecting COVID-19-infected patients (Lin Li et al. 2020; Salman et al. 2020). As compared to DLMs, TLMs outperform in terms of detection accuracy, generalization error, and training time at the cost of a vast amount of labeled training data (Salman et al. 2020).

This chapter focuses on the development of a computationally efficient automated detection scheme by utilizing modified Generative Adversarial Network (MGAN)–based transfer learning (TL) for the successful detection of COVID-19 cases using chest X-ray images. Initially, the MGAN model is used for the generation of a synthetic dataset with varied feature attributes for the initial training of DLMs. MGAN is more computationally efficient than the usual GAN model, as it utilizes a lighter weight network architecture instead of a deep network, and is free from the problem of instability by utilizing a weighted combination of content and structural loss functions. Consequently, the MGAN model produces realistic generated samples at a faster convergence speed. Afterward, various benchmark DLMs such as VGG19, ResNet50, InspectionV3, InspectionResnetV2, DenseNet121, DenseNet169, DenseNet201 and MobileNet are fine-tuned with the generated dataset. In the testing phase, the COVID-19 infected X-ray images are fed to these DLMs for the purpose of classification and an outperforming model is chosen for real-time use. Extensive experimental analyses have been performed on popularly available image datasets. The detection accuracy of our work is compared with similar state-of-the-art detection methods. The suggested MGAN model along with the pre-trained MobileNet model outperforms the other models.

The discussion flow of this chapter is structured in four sections. Section 3.2 delivers a brief discussion on some of the deep learning (DL)–based COVID-19 disease detection methods by using X-ray images in literature. Section 3.3 discusses the MGAN model for the generation of a dataset for the training of DLMs. Section 3.4 focuses on the performance analysis of our suggested work compared with some of the similar classification methods in the literature. Section 3.5 concludes the chapter, along with some future scope for this work.

3.2 LITERATURE SURVEY

AAI-based methods utilizing various radiometric images (chest CT scan and X-ray images) provide an excellent breakthrough in detecting COVID-19-infected persons. This section briefly summarizes some of the existing DL-based models (Rahul Kumar et al. 2020; Linda Wang and Alexander Wong 2020; P. K. Sethy and S. K. Behera 2020; Ali Narin et al. 2020; Tulin Ozturk et al. 2020; Karim Hammoudi et al. 2020; I.D. Apostolopoulos et al. 2020; E.E.D. Hemdan et al. 2020; Abdul Waheed et al. 2020; N.E.M. Khalifa et al. 2020; Mohamed Loey et al. 2020; Parnian Afshar et al. 2020; Ferhat Ucar and Korkmaz Deniz 2020; M.Z. Islam et al. 2020; T. Mahmud et al. 2020) by utilizing chest X-ray images for the detection of COVID-19 infections from uninfected cases. Table 3.1 reviews the brief description of these

TABLE 3.1
COVID-19 Infection Diagnosis via Existing DLMs Using Chest X-Ray Images

References	Number of Cases Used in the Experimentation	Used DL Model	Observations (Accuracy)
Rahul Kumar et al. 2020	Training set: (42 COVID-19 cases + 894 normal + 897 pneumonia cases). Testing set: (20 COVID-19 + 447 normal + 448 pneumonia cases)	ResNet152 model + SMOTE	Accuracy using Random Forest: 0.973 Using XGBoost predictive classifier: 0.977
Linda Wang and Alexander Wong 2020	53 COVID-19 + 5,526 Non-COVID-19 cases + 8,066 healthy cases	COVID-Net	Model helps the clinicians to screen patients with an accuracy of 0.924
P.K. Sethy and S.K. Behera 2020	50 COVID-19 cases + 50 Non-COVID-19 cases	ResNet50 + SVM	Classification accuracy via SVM is 0.953
Ali Narin et al. 2020	50 COVID-19 cases + 50 Non-COVID-19 cases	Deep CNN + ResNet-50	Accuracy for binary class classification is 0.98
Tulin Ozturk et al. 2020	125 COVID-19 cases + 500 Non-COVID + 500 pneumonia cases	DarkNet + 17 layered model	For binary class, accuracy is 0.0980 and, for multi-class detection, accuracy is 0.870
Karim Hammoudi et al. 2020	Training set: 5,863 images Test set: 145 COVID-19 cases	Tailored DL + ResNet34, VGG16 ResNet50 and DenseNet169	Classification accuracies for COVID vs bacteria is 0.979
I.D. Apostolopoulos et al. 2020	224 COVID-19 + 700 pneumonia cases + 504 healthy cases	TL + VGG19 model	Model achieves accuracy of 0.9875 for binary classification
E.E.D. Hemdan et al. 2020	25 COVID-19 cases + 25 normal cases	COVIDX-Net	The model detects at an accuracy of 0.90
Abdul Waheed et al. 2020	Generated via GAN: 1,399 normal cases + 1,669 COVID-19 cases	GAN + VGG16 model	The model achieves a classification accuracy of around 0.95
N.E.M. Khalifa et al. 2020	Generated via GAN: Normal cases: 5,863 images COVID-19: 6,240 cases	GAN + AlexNet, GoogLeNet, SqueezeNet, Resnet18	Model using Resnet18 achieved a classification accuracy of 0.99
Mohamed Loey et al. 2020	Generated via GAN: Normal cases: 2,100 images COVID-19 cases: 1,800 images	GAN + AlexNet	Classification accuracy is around 100% for the binary class problem

(Continued)

TABLE 3.1 (CONTINUED)
COVID-19 Infection Diagnosis via Existing DLMs Using Chest X-Ray Images

References	Number of Cases Used in the Experimentation	Used DL Model	Observations (Accuracy)
Parnian Afshar et al. 2020	COVID-19 cases: 53 images Non-COVID-19 cases: 5,526	CapsNet	CapsNet model obtains classification accuracy of 0.9830
Ferhat Ucar and Korkmaz Deniz 2020	Total augmented cases: 4,608 COVID-19: pneumonia: normal: 1,536: 1,536: 1,536 images	Bayes-SqueezeNet	Classification accuracy for the three-class problem is 0.9830
M.Z. Islam et al. 2020	Total cases: 4,575 images COVID-19: normal: pneumonia cases are 1,525: 1,525: 1,525	CNN+LSTM	Detection accuracy for two-class problem is around 0.994
T. Mahmud et al. 2020	Total cases: 5,856 images Normal cases: 1,583 Non-COVID: 1,493 Bacterial pneumonia: 2,780	CovXNet	Pre-trained CovXNet model provides detection accuracy of 0.974 for two-class problem

classification models along with the size of the dataset used and achieved classification accuracies for different class problems.

Rahul Kumar et al. (2020) used both ML and DL techniques for the accurate prediction of COVID-19 patients. Here, a ResNet152 model with synthetic minority over-sampling technique (SMOTE) was used to extract deep features, whereas random forest and XGBoost classifiers are used for the binary classification. Linda Wang and Alexander Wong (2020) have proposed the COVID-Net model to classify the COVID-19 images with an accuracy of 92.4%. P. K. Sethy and S. K. Behera (2020) utilized various CNN models to extract features of the X-ray images, and these feature vectors are classified via support vector machine (SVM) classifier. They found that ResNet50 model with SVM classifier provides better accuracy. Ali Narin et al. (2020) used the pre-trained ResNet50 model for the classification purpose. Tulin Ozturk et al. (2020) utilized the pre-trained DarkNet model for the binary and multi-class classification of COVID-19 images. Karim Hammoudi et al. (2020) have proposed a tailored deep learning model via various CNNs architectures such as ResNet34, ResNet50, DenseNet169, VGG19 and InceptionResNetV2–RNN to detect multi-class pneumonia-infected cases using chest X-ray images. I.D. Apostolopoulos et al. (2020) utilized a TL-based model with pre-trained VGG19 model for the classification purpose. E.E.D. Hemdan et al. (2020) utilized a DL-based model named COVIDX-Net, comprising of seven CNN models to diagnose COVID-19 X-ray images. Abdul Waheed et al. (2020), N.E.M. Khalifa et al. (2020), and Mohamed Loey et al. (2020) utilized a Generative Adversarial Network (GAN) model in the data augmentation process to generate a sufficient amount of training images, followed by the utilization of various TL models for the classification process. TL models pre-trained via the generated images boost up the classification accuracy. Parnian Afshar et al. (2020) and Ferhat Ucar and Korkmaz Deniz (2020) utilized capsule network-based and deep Bayes-SqueezeNet-based frameworks, respectively, for the identification of COVID-19 cases. In contrast, M.Z. Islam et al. (2020) utilized the combination of deep CNN framework with the long short-term memory network for the detection of the novel coronavirus. Deep features were extracted via CNN, and these extracted features were fed to the later model for the classification process. T. Mahmud et al. (2020) utilized multi-receptive features by utilizing multi-dilation CNN models for the automatic detection of COVID-19 infections.

3.3 PROPOSED METHOD

Performance accuracy of DL-based classification methodology is severely influenced by the size of the training dataset required for the fine-tuning of parameters. The novel COVID-19 epidemic is recent. Hence, the collection of vast amounts of COVID-19 infected radiometric images are quite difficult. However, DLMs are data-hungry models; hence, the classification process may overfit due to a lack of sufficient data. As a solution, some of the existing methods (Linda Wang and Alexander Wong 2020; P. K. Sethy and S. K. Behera 2020; Ali Narin et al. 2020; E.E.D. Hemdan et al. 2020) utilized images in the *ImageNet* (Jia Deng et al. 2009) dataset to fine-tune the parameters. *ImageNet* dataset consists of several millions of natural images whose characteristics are quite different from the COVID-19 infected radiometric images. Hence, the pre-training of DLMs via the *ImageNet* dataset produces inaccurate detection with a higher false rate. Methods reported by Abdul Waheed et al. (2020), Mohamed Loey et al. (2020), M.Z. Islam et al. (2020), and T. Mahmud et al. (2020) utilized data augmentation techniques to expand the size of the training dataset synthetically. Typical data augmentation schemes are comparatively faster and easier to perform. However, these schemes fail to produce new and informative training samples with varied feature attributes. In contrast, methods reported by Abdul Waheed et al. (2020), Mohamed Loey et al. (2020), Parnian Afshar et al. (2020), and T. Mahmud et al. (2020) utilized an external dataset consisting of radiometric images of other types of viral infections such as pneumonia and viral influenza for training the DLMs. Methods reported by Abdul Waheed et al. (2020), N.E.M. Khalifa et al. (2020), and Mohamed Loey et al. (2020) utilized the Generative Adversarial Network (GAN) model as the pre-processing step to synthetically boost up the volume of the training dataset. GAN is an influential alternative to yield realistic results but some of the inherent bottlenecks such as (i) computational overload, (ii) internal instability, (iii) mode collapse and (iv) convergence problem may limit its use in large-scale applications.

To deal with the above issues, our proposed scheme utilizes GAN-based TL for the successful classification of COVID-19 cases. The entire classification process is carried out in three successive steps: *step 1*: data collection and augmentation;

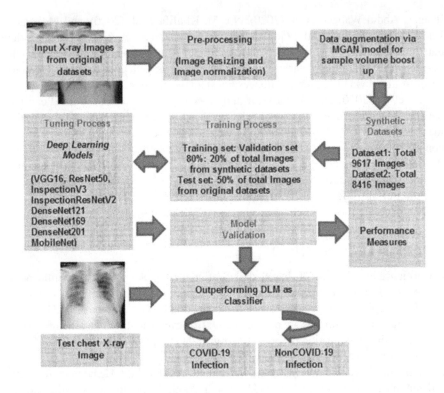

FIGURE 3.1 Schematic workflow of the proposed work.

step 2: generation of synthetic COVID-19 X-ray image dataset by utilizing modified GAN (MGAN) model; *step 3*: fine-tuning of TLMs for the classification purpose. Extensive experimental analyses have been performed on the popularly available COVID-19 X-ray image datasets as well as synthetically generated datasets to assess the accuracy of our proposed scheme. The detection accuracy of our work is compared with similar state-of-the-art detection methods. Figure 3.1 depicts the schematic workflow of our suggested method.

3.3.1 Data Collection and Preprocessing

Here, we have collected sample COVID-19 and non-COVID19 infected X-ray images from publicly available datasets (J.P. Cohen et al. 2020; Dataset 2020). The collected images are of varied resolution and varied sizes. Hence, to maintain uniformity and to speed up the testing process, a few preprocessing operations such as image resizing and image normalization are performed on the collected data.

3.3.2 Generation of Expanded Dataset via MGAN

This step aims at the expansion of samples in Dataset 1 and Dataset 2 synthetically via the MGAN model. GAN is one of the state-of-the-art models to generate complete unseen images synthetically by performing unsupervised min-max optimization. GAN utilizes two contrasting networks such as the generator network (GN) and the discriminator network (DN) comprising several residual blocks (RBs) in the network. Conventionally, convolutional layers usually consist of 3 × 3 sized kernels and 128 feature maps followed by the utilization of batch normalization layers (BNLs). The input/real X-ray image (I_X) is fed to the GN and the network aims at generating a realistic image (I_{Gen}), whose probability is quite similar to that of the original ground-truth image, by minimizing the loss function between the generated and original images. Further, either I_{Gen} or I_X image is fed to the DN which aims at distinguishing the real one from the generated one efficiently by maximizing the probability of distribution. GN is parameterized as $\left(GN_{\theta_{GE}}\right)$ and DN is parameterized as $\left(DN_{\theta_{DI}}\right)$, respectively. In a nutshell, the whole model is characterized as an adversarial min-max optimization problem (B.Z. Demiray et al. 2020) as described in equation (3.1). Objective function of the whole adversarial process is described in equation (3.2).

$$\min_{\theta_{GE}} \max_{\theta_{DI}} \mathrm{E}_{I_X \sim P_{Train}(I_X)}\left[\log DN_{\theta_{DI}}\left(I_X\right)\right]$$
$$+ \mathrm{E}_{I_{Gen} \sim P_{GN}(I_{Gen})}\left[\log\left(1 - DN_{\theta_{DI}}\left(GN_{\theta_{GE}}\left(I_X\right)\right)\right)\right] \quad (3.1)$$

$$\mathrm{Obj}(\theta_{GE}, \theta_{DI}) = loss_{GN} + \lambda \; loss_{DI}(\theta_{GE}, \theta_{DI}) \quad (3.2)$$

where $loss_{GN}$ is the loss function associated with the GN whose sole purpose is to improve the image generation quality and to minimize the effect of artifacts in the generated images. $loss_{DI}$ denotes the adversarial loss associated with the DN. Inconsistent choice of these loss functions will generate internal instability in the model and may generate low-quality output X-ray images or may get stuck in local minima leading to mode collapse problem. Volume of convolutional layers and BNLs contribute toward the computational burden of the optimization process.

Unlike the conventional GAN model, MGAN is computationally efficient by utilizing lighter weight network architecture instead of deep network. GN of the MGAN model replaces the bulky RBs with residual-in-residual-dense-blocks (RRDBs) (X. Wang et al. 2018). RRDBs exploit the benefit of dense connections and multilevel residual architecture, whereas in DN architecture BNLs are completely removed (B. Lim et al. 2017). Removing BNLs will help us to deal with the feature range flexibility with a drastic reduction in memory usages. The MGAN model utilized a weighted combination of content loss, texture loss, and diversified features captured at different stages of RRDBs interpretations. Consequently, the MGAN model converges at a faster rate, generates high-quality realistic X-ray images, and improves the internal stability of the entire process than the conventional GAN model. Figure 3.2 depicts the architecture of GN and DN in the MGAN model by replacing the conventional residual blocks with the RRDBs and by removing BNLs.

Equation (3.3) describes the modified formulation of the objective function $loss_{Total}(\theta_{GE}, \theta_{DI})$ to be utilized in the optimization process.

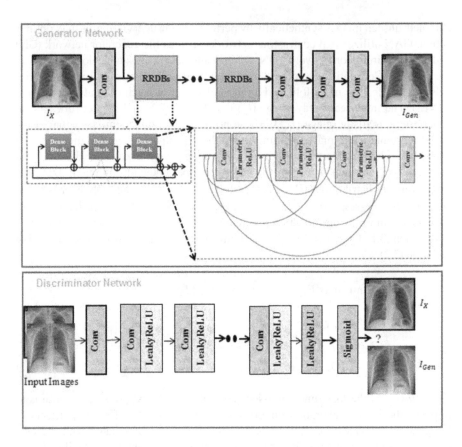

FIGURE 3.2 MGAN model architecture: (a) GN by replacing RBs with RRDBs; (b) DN by removing BNLs.

$$loss_{Total}(\theta_{GE}, \theta_{DI}) = loss_{Texcont}(\theta_{GE}) + \alpha\, loss_{DN}(\theta_{GE}, \theta_{DI}) + \beta\, loss_{GDF}(\theta_{GE})$$

$$loss_{Texcont}(\theta_{GE}) = \left\| I_{Gen} - I_X \right\|_2 + \lambda \left\| MsTex(I_{Gen}) - MsTex(I_X) \right\|_2$$

(3.3)

$$loss_{DN}(\theta_{GE}, \theta_{DI}) = -\log DN(I_X) - \log(1 - DN(GN(I_{Gen})))$$

$$loss_{GDF}(\theta_{GE}) = \left\| GDF(I_{Gen}) - GDF(I_X) \right\|_2$$

where I_X is the original X-ray image; I_{Gen} is the generated image by the GN; α, β, and λ are the weight controlling parameters and their values are chosen as 0.001, 4, and 0.01 respectively. $loss_{Texcont}$ and $loss_{GDF}$ are used to train the GN, whereas $loss_{DN}$ is used to train the DN. $loss_{Texcont}$ integrates a weighted combination of content loss and multi-scale textural loss. Content loss is computed as the pixel-wise mean square error (MSE) loss. $MsTex(.)$ function finds the multi-scale textural information of the image under consideration which is obtained by convolving the image with the group

TABLE 3.2
Implementation Details of the MGAN Model

Generator Network		Discriminator Network	
Parameters	Values	Parameters	Values
Conv kernel size	(3×3)	Conv kernel size	(3×3)
Conv feature map	64, 128, 256	Conv feature map	128
Activation function	ParametricReLU	Activation function and final activation function	LeakyReLU type/ Sigmoid type
Learning rate	0.1	Learning rate	0.2
Number of RRDBs	5	Number of Conv layers	8
Resized input image size	(224×224)	Optimizer	Adam function
Momentum, exponential decay rate of Adam optimizer	0.5, 0.5	Number of epochs	50

of Gabor filter banks (R. Nayak and D. Patra 2018). Gabor filter banks efficiently preserve sufficient high-frequency details and textural content of the image along multi-scale $sc = \{1,2\}$ and multi-orientations $ori = \left\{0, \frac{\pi}{4}, \frac{\pi}{2}, \pi\right\}$. Moreover, these filter banks are good at preserving several salient visual characteristics such as directional features and spatial position of the image. $lossGDF(.)$ function is used to measure the discrepancy between the global dense feature (GDF) of the real and generated X-ray images. GDF of an image is computed by fusing the feature from all the RRDBs used in the GN and is described as $GDF = [F_1, F_2, ..., F_n]$. $[F_1, F_2, ..., F_n]$ corresponds to the concatenations of feature maps extracted from n number RRDBs used in the GN. As a result, diversified features at different stages of RRDB interpretations are properly apprehended in the optimization process. Iterative execution of the optimization process helps to boost up the volume of images present in the publicly available datasets by 30 times. Table 3.2 lists the parameters used in the proposed MGAN model. Table 3.3 provides the detailed constituents of original and synthetically generated Dataset 1 and Dataset 2.

3.3.3 Fine-tuning Transfer Learning Models for Classification

In this section, various DLMs such as VGG19 (Karen Simonyan and Andrew Zisserman 2014), ResNet50 (K. He et al. 2016), InspectionV3 (C. Szegedy et al. 2016), InspectionResnetV2 (C. Szegedy and S. Ioffe et al. 2016), DenseNets (121, 169, 201) (G. Huang et al. 2017), and MobileNet (G.A. Howard et al. 2017) models are pre-trained to fine-tune their hyperparameters. These DLMs are trained and validated with a larger amount of images in synthetically generated Dataset 1 and Dataset 2. Eighty percent of the total images are used for training and 20% are used for validation purposes. After training of DLMs, validation is performed to select the best performing DLM to be integrated into the TL approach

TABLE 3.3
Detailed Constituents of Original and Synthetically Generated Dataset 1 and Dataset 2

	Original Dataset 1				Original Dataset 2		
Type	COVID-19 Images	Non-COVID-19 Images	Total	Type	COVID-19 Images	Non-COVID-19 Images	Total
Training	73	155	228	Training	45	154	199
Validation	28	60	88	Validation	17	59	76
Testing	11	24	35	Testing	7	24	31
Total	112	239	351	Total	69	237	306
Synthetically generated Dataset 1				*Synthetically generated Dataset 2*			
Training	2,184	4,661	6,809	Training	1,346	4,622	5,968
Validation	840	1,792	2,632	Validation	517	1,777	2,294
Testing	56	120	176	Testing	35	119	154
Total	3,080	6,573	9,617	Total	1,898	6,518	8,416

TABLE 3.4
Implementation Details of DLMs

Parameter	VGG19	ResNet50	InceptionV3	InceptionResNetV2	DenseNet	MobileNet
Optimizer	RMSProp	RMSProp	SGD	RMSProp	Adam	RMSProp
Input size	(224×224)	(224×224)	(299×299)	(299×299)	(224×224)	(224×224)

for classifying COVID-19 patients. Fifty percent of the total images from the original datasets are used as testing dataset. In the training process, the top layers of the models are frozen, whereas the bottom layers are only trained. This is because the bottom layers are efficient enough to preserve sufficient image details. Consequently, the freezing process allows the reduction of computational burden without hampering the classification accuracy. Performance behavior of each model has been recorded in terms of training and validation accuracy as well as training and validation loss. All these DLMs employ a common loss function as binary cross-entropy and are iterated for 50 epochs. Table 3.4 shows the implementation details of the used DLMs.

3.4 RESULT AND ANALYSIS

This section discusses the performance analysis of our proposed work. Performance analysis is accomplished in two steps: (i) selection of outperforming DLM; (ii) performance comparison of our proposed TL-based approach using the outperforming DLM (from step 1) with some of the identical classification methods (Abdul Waheed et al. 2020; N.E.M. Khalifa et al. 2020; Mohamed Loey et al. 2020) found in literature. The proposed work utilizes Keras (C. Szegedy and S. Ioffe et al. 2016) as the key deep learning framework and TensorFlow is used as the back-end engine. Besides, several libraries such as Scikit-learn, sklearn, PIL, Pandas, and NumPy are also employed for implementing the work. A computer server with a dedicated GPU with AMD Ryzen 53550H processor and Nvidia GeForce GTX 1050 (Dataset 2020) graphics card is used. Classification assessment is performed by using X-ray images in Dataset 1 and Dataset 2.

Selection of outperforming DLM to be embedded in the TL approach is decided by the validation accuracy of the models. Table 3.5 depicts the average training accuracy, training loss, validation accuracy, and validation loss of the used DLMs. Best outcomes are represented in bold faces. From the tabular data analysis, it is observed that the MobileNet model outperforms the other models in terms of training and validation accuracy. This model achieves 99.97% training accuracy and 99.34% validation accuracy. These accuracies seem to be worthy enough for the classification of COVID-19 cases from the sample X-ray images. Nevertheless, training and validation loss of this model is sufficient enough to prevent extreme cases of underfitting or overfitting problems. Figure 3.3 plots average validation loss with respect to the number of epochs. From the figure, it is observed that the validation loss curve of all

TABLE 3.5
Performance Analysis of the Pre-trained DLMs on Generated Dataset 1 and Dataset 2

Model	Avg. Training Accuracy	Avg. Training Loss	Avg. Validation Accuracy	Avg. Validation Loss
VGG19	0.9928	0.0334	0.9286	0.2324
ResNet50	0.9979	0.0062	0.8857	3.1410
InspectionV3	0.9033	0.6725	0.8257	4.2745
InspectionResNetV2	0.9175	0.6317	0.8457	4.0151
DenseNet121	0.9726	0.0865	0.8957	2.2245
DenseNet169	0.9750	0.0778	0.9088	1.3131
DenseNet201	0.9839	0.0602	0.8954	4.7131
MobileNet	**0.9997**	**0.0013**	**0.9934**	**0.0016**

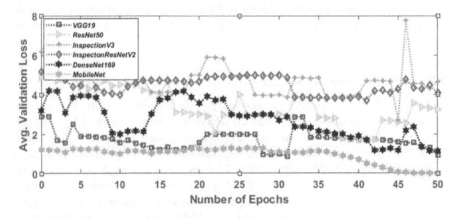

FIGURE 3.3 Validation loss curves of the used DLMs.

the DLMs except VGG19 and MobileNet model increases at the beginning and progresses with ups and downs with an increase in the number of epochs. For VGG19 and MobileNet model, the loss curves increase initially and gradually decrease toward the end of the iteration. Moreover, when compared to the VGG19 model, MobileNet exhibits lower validation loss. Hence, the MobileNet model is selected as one of the outperforming DLM to be included in the TL approach for the classification process. Our proposed TL model utilizing the MGAN model and pre-trained MobileNet model is further referred to as the *MoBMGAN* model.

The second part of the analysis focuses on the performance assessment of our proposed *MoBMGAN* model with some of the existing classification methods reported by Abdul Waheed et al. (2020), N.E.M. Khalifa et al. (2020), and Mohamed Loey et al. (2020). In the classification process, true predictions are characterized by True Positive Rate (TPR) and True Negatives Rate (TNR), whereas false predictions are characterized by False Positives Rate (FPR) and False Negative Rate (FNR). Performance behavior of our proposed model along with compared models is characterized by various quality measures such as Accuracy, Precision, Recall, IoU score, and F_1 score, by utilizing these true and false prediction results. Computational complexity and memory usage of these models are characterized by the parameters FLOPs and GPU memory, respectively. Equation (3.4) represents the formulation of these quality assessment parameters. Higher values of Accuracy, Precision, Recall, IoU score, and F_1 score and lower values of FLOPs and GPU memory signifies superior classification accuracy. IoU score parameter (M.A. Rahman et al. 2016) evaluates the accuracy of a trained classifier by measuring the match between predicted and real images. FLOPs (X. Xie et al. 2020) stands for floating-point operations which are used to characterize the computational burden of any model by counting the number of multiplication and addition operations required for the execution of the model. GPU memory parameter is used to measure the complexity of the model in terms of the size of memory requirement (in MB) to store the learnable parameters. Table 3.6 summarizes the above-mentioned parameters of our proposed *MoBMGAN* model along with the compared methods for Dataset 1 and Dataset 2 images.

TABLE 3.6
Performance Assessment of Our Proposed Model Compared with Existing Methods

Models	Accuracy	Precision	Recall	IoU score	F_1 score	FLOPs (million)	GPU Memory (million)
				Dataset 1			
[12]	0.9546	0.9012	0.9123	0.9034	0.9067	31.909	137.170
[13]	0.9900	0.9154	0.9321	0.9110	0.9236	23.456	122.321
[14]	0.9923	0.9223	0.9430	0.9219	0.9326	**73.4512**	**249.000**
MoBMGAN (Proposed)	**0.9967**	**0.9316**	**0.9567**	**0.9330**	**0.9440**	16.4249	105.853
				Dataset 2			
[12]	0.9523	0.9045	0.9137	0.9055	0.9091	31.909	137.170
[13]	0.9903	0.9166	0.9342	0.9179	0.9253	23.456	122.321
[14]	0.9962	0.9287	0.9463	0.9255	0.9374	**73.4512**	**249.000**
MoBMGAN (Proposed)	**0.9975**	**0.9380**	**0.9574**	**0.9387**	**0.9476**	16.4249	105.853

Note: Higher values of Accuracy, F1 Score and IoU score signify better classification capability of the model. Whereas, lower values of FLOPs and GPU Memory signify lower computational complexity and minimal memory usage of the used models.

$$\text{Accuracy} = \frac{(TPR + TNR)}{(TPR + TNR + FPR + FNR)}; \text{Recall} = \frac{TPR}{(TPR + FNR)};$$

$$\text{Precision} = \frac{TPR}{(TPR + FPR)}; \quad F_1 \text{ score} = \frac{2(\text{Precision} * \text{Recall})}{(\text{Precision} + \text{Recall})}; \quad (3.4)$$

$$\text{IoU score} = \frac{TPR}{(TPR + FPR + FNR)}$$

From the tabular data, it is observed that the method reported by Abdul Waheed et al. (2020) achieves the least classification accuracy of around 95.35% than the other compared methods. This is because VGG19 architecture trained via training images generated via the conventional GAN model cannot provide ultimate accuracy in the classification process. In contrast, N.E.M. Khalifa et al. (2020) utilized GAN for generating enough initial training images followed by the integration of a limited layered CNN architectures-based TL model to achieve a classification accuracy of 99.02%.

Likewise, the method reported by Mohamed Loey et al. (2020) utilized GAN to boost up the volume of training samples by 30 times and utilized the TL model using pre-trained AlexNet architecture to achieve the classification accuracy of around 99.43%. When compared to these methodologies, our proposed *MoBMGAN* model utilizes MobileNet architecture, pre-trained via sufficient high-quality initial training images and generated by the MGAN model. This enables one to preserve sufficient image features in the classification process and consequently attain an average classification accuracy of 99.71%. Our proposed model outperforms not only in terms of classification accuracy but also in terms of other quality measures. Figure 3.4 depicts the comparison plot of these parameters. Figure 3.5 shows some of the true and false predictions of the compared models along with our proposed model.

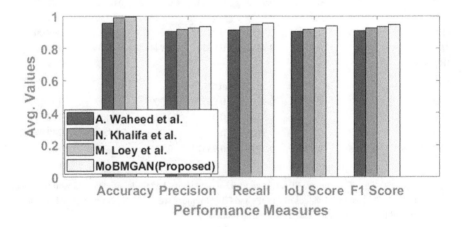

FIGURE 3.4 Performance assessments of compared methods.

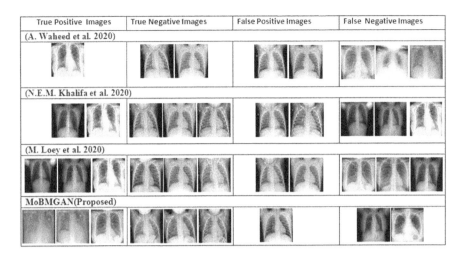

FIGURE 3.5 Correct and incorrect predictions of COVID-19 cases using X-ray images.

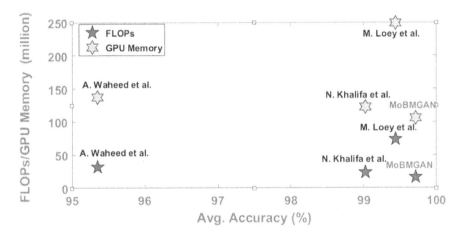

FIGURE 3.6 Accuracy vs FLOPs and GPU memory of compared models. *MoBMGAN* is our proposed model.

3.4.1 ACCURACY VS COMPUTATIONAL COMPLEXITY

Parameters FLOPs and GPU memory requirements characterize the computational complexity of any model. Figure 3.6 shows the plot of average accuracy to FLOPs (millions) and GPU memory (millions) requirement of the compared methods. From the given plot, it is observed that our proposed work provides an excellent trade-off between the classification accuracy to computational complexity. The lower computational burden of our work is due to the integration of RRDBs instead of bulky RBs in the GAN model as well as the utilization of depth-wise separable-convolution modules of the MobileNet model. However, it is also observed that the method reported by Mohamed Loey et al. (2020) provides comparable classification accuracy at the cost of larger computational overhead. This is due to the integration of a computationally complex GAN model along with AlexNet architecture. The TL approach using the AlexNet model offers complex architecture and thus requires larger space or larger GPU requirements to store the parameters. In contrast, our proposed *MoBMGAN* model using MobileNets is compact and computationally efficient. Nevertheless, it gives a state-of-the-art performance in providing a superior rate of accuracy and lowers false positives for the automatic detection of COVID-19 cases using X-ray images.

3.5 CONCLUSION

Transfer learning (TL) based approaches including different pre-trained DLMs afford outstanding breakthroughs in the classification of COVID-19 cases from non-COVID cases by utilizing radiometric images such as CT scans and X-ray images. However, inadequate and disparity of training images in the COVID-19 dataset affect a lot to the performance behavior of TL-based classification models. This work proposed a TL-based method by employing various pre-trained DLMs such as VGG19, ResNet50, InspectionV3, InspectionResNetV2, DenseNets, and MobileNet to screen out COVID-19 patients by using chest X-ray images. These DLMs are pre-trained with COVID-19 infected X-ray image datasets consisting of a larger volume of data samples, generated artificially via a modified GAN model. Consequently, these models do not suffer from over-fitting or under-fitting problem. Among these pre-trained DLMs, MobileNet outperforms the other models in terms of classification accuracy and convergence speed. The proposed TL-based approach utilizing MGAN and pre-trained MobileNet model successfully assists medical professionals in efficiently screening out COVID-19 infected patients and taking early diagnostic decisions.

In the future, hybridization of low-cost deep learning models along with optimized features attributes obtained via various optimization algorithms will enable us to distinguish COVID-19 patients with mild symptoms from other similar viral infections at greater accuracy. Moreover, the future direction of the work will focus on devolving intelligent techniques for the detection of COVID-19 infections by multi-modal fusion of various radiometric images.

REFERENCES

Afshar, Parnian, Shahin Heidarian, Farnoosh Naderkhani, Anastasia Oikonomou, Konstantinos N. Plataniotis, and Arash Mohammadi. "Covid-caps: A capsule network-based framework for identification of covid-19 cases from x-ray images." *arXiv Preprint arXiv:2004.02696* (2020).

Apostolopoulos, Ioannis D., and Tzani A. Mpesiana. "Covid-19: Automatic detection from x-ray images utilizing transfer learning with convolutional neural networks." *Physical and Engineering Sciences in Medicine* 43(2020): 635–640.

Cohen, Joseph Paul, Paul Morrison, Lan Dao, Karsten Roth, Tim Q. Duong, and Marzyeh Ghassemi. "Covid-19 image data collection: Prospective predictions are the future." *arXiv Preprint arXiv:2006.11988* (2020).

Dataset. Available online: https://drive.google.com/uc?id=1coM7x3378f-Ou2l6Pg2wlda
Ol7Dntu1a (accessed on 31 March 2020).

Demiray, Bekir Z., Muhammed Sit, and Ibrahim Demir. "D-SRGAN: DEM super-resolution with generative adversarial network." *arXiv Preprint arXiv:2004.04788* (2020).

Deng, Jia, Wei Dong, Richard Socher, Li-Jia Li, Kai Li, and Li Fei-Fei. "Imagenet: A large-scale hierarchical image database." In *2009 IEEE Conference on Computer Vision and Pattern Recognition* (2009), IEEE, pp. 248–255.

Hammoudi, Karim, Halim Benhabiles, Mahmoud Melkemi, Fadi Dornaika, Ignacio Arganda-Carreras, Dominique Collard, and Arnaud Scherpereel. "Deep learning on chest X-ray images to detect and evaluate pneumonia cases at the era of COVID-19." *arXiv Preprint arXiv:2004.03399* (2020).

He, Kaiming, Xiangyu Zhang, Shaoqing Ren, and Jian Sun. "Deep residual learning for image recognition." In *Proceedings of the IEEE Conference on Computer Vision and Pattern Recognition* (2016), pp. 770–778.

Hemdan, Ezz El-Din, Marwa A. Shouman, and Mohamed Esmail Karar. "Covidx-net: A framework of deep learning classifiers to diagnose covid-19 in x-ray images." *arXiv Preprint arXiv:2003.11055* (2020).

Howard, Andrew G., Menglong Zhu, Bo Chen, Dmitry Kalenichenko, Weijun Wang, Tobias Weyand, Marco Andreetto, and Hartwig Adam. "Mobilenets: Efficient Convolutional Neural Networks for Mobile Vision Applications." *arXiv Preprint arXiv:1704.04861* (2017).

Huang, Gao, Zhuang Liu, Laurens Van Der Maaten, and Kilian Q. Weinberger. "Densely connected convolutional networks." In *Proceedings of the IEEE Conference on Computer Vision and Pattern Recognition* (2017), pp. 4700–4708.

Islam, Md Zabirul, Md Milon Islam, and Amanullah Asraf. "A combined deep CNN-LSTM network for the detection of novel coronavirus (COVID-19) using X-ray images." *Informatics in Medicine Unlocked* 20 (2020): 100412.

Khalifa, Nour Eldeen M., Mohamed Hamed N. Taha, Aboul Ella Hassanien, and Sally Elghamrawy. "Detection of coronavirus (COVID-19) associated pneumonia based on generative adversarial networks and a fine-tuned deep transfer learning model using chest X-ray dataset." *arXiv Preprint arXiv:2004.01184* (2020).

Kumar, Rahul, Ridhi Arora, Vipul Bansal, Vinodh J. Sahayasheela, Himanshu Buckchash, Javed Imran, Narayanan Narayanan, Ganesh N. Pandian, and Balasubramanian Raman. "Accurate prediction of COVID-19 using chest X-ray Images through deep feature learning model with SMOTE and machine learning classifiers." *medRxiv* (2020).

Li, Lin, Lixin Qin, Zeguo Xu, Youbing Yin, Xin Wang, Bin Kong, Junjie Bai et al. "Artificial intelligence distinguishes COVID-19 from community acquired pneumonia on chest CT." *Radiology* (2020).

Lim, Bee, Sanghyun Son, Heewon Kim, Seungjun Nah, and Kyoung Mu Lee. "Enhanced deep residual networks for single image super-resolution." In *Proceedings of the IEEE Conference on Computer Vision and Pattern Recognition Workshops* (2017), pp. 136–144.

Loey, Mohamed, Florentin Smarandache, and Nour Eldeen M Khalifa. "Within the lack of chest COVID-19 X-ray dataset: A novel detection model based on GAN and deep transfer learning." *Symmetry* 12, no. 4 (2020): 651.

Mahmud, Tanvir, Md Awsafur Rahman, and Shaikh Anowarul Fattah. "CovXNet: A multi-dilation convolutional neural network for automatic COVID-19 and other pneumonia detection from chest X-ray images with transferable multi-receptive feature optimization." *Computers in Biology and Medicine* 122 (2020): 103869.

Narin, Ali, Ceren Kaya, and Ziynet Pamuk. "Automatic detection of coronavirus disease (covid-19) using X-ray images and deep convolutional neural networks." *arXiv Preprint arXiv:2003.10849* (2020).

Nayak, Rajashree, and Dipti Patra. "Enhanced iterative back-projection based super-resolution reconstruction of digital images." *Arabian Journal for Science and Engineering* 43, no. 12 (2018): 7521–7547.

Ozturk, Tulin, Muhammed Talo, Eylul Azra Yildirim, Ulas Baran Baloglu, Ozal Yildirim, and U. Rajendra Acharya. "Automated detection of COVID-19 cases using deep neural networks with X-ray images." *Computers in Biology and Medicine* 121(2020): 103792.

Rahman, Md Atiqur, and Yang Wang. "Optimizing intersection-over-union in deep neural networks for image segmentation." In International Symposium on Visual Computing (2016), Springer, Cham, pp. 234–244.

Salman, Fatima M., Samy S. Abu-Naser, Eman Alajrami, Bassem S. Abu-Nasser, and Belal A.M. Alashqar. *COVID-19 Detection Using Artificial Intelligence.* (2020).

Sethy, Prabira Kumar, and Santi Kumari Behera. "Detection of coronavirus disease (covid-19) based on deep features." *Preprints* 2020030300 (2020): 2020.

Simonyan, Karen, and Andrew Zisserman. "Very deep convolutional networks for large-scale image recognition." *arXiv Preprint arXiv:1409.1556* (2014).

Szegedy, Christian, Sergey Ioffe, Vincent Vanhoucke, and Alex Alemi. "Inception-v4, inception-resnet and the impact of residual connections on learning." *arXiv Preprint arXiv:1602.07261* (2016).

Szegedy, Christian, Vincent Vanhoucke, Sergey Ioffe, Jon Shlens, and Zbigniew Wojna. "Rethinking the inception architecture for computer vision." In *Proceedings of the IEEE Conference on Computer Vision and Pattern Recognition* (2016), pp. 2818–2826.

Ucar, Ferhat, and Deniz Korkmaz. "COVIDiagnosis-Net: Deep Bayes-squeeze net based diagnostic of the Coronavirus Disease 2019 (COVID-19) from X-ray images." *Medical Hypotheses* 140(2020): 109761.

Waheed, Abdul, Muskan Goyal, Deepak Gupta, Ashish Khanna, Fadi Al-Turjman, and Plácido Rogerio Pinheiro. "COVIDGAN: Data augmentation using auxiliary classifier gan for improved covid-19 detection." *IEEE Access* 8 (2020): 91916–91923.

Wang, Linda, and Alexander Wong. "COVID-Net: A tailored deep convolutional neural network design for detection of COVID-19 cases from chest X-ray images." *arXiv Preprint arXiv:2003.09871* (2020).

Wang, Xintao, Ke Yu, Shixiang Wu, Jinjin Gu, Yihao Liu, Chao Dong, Yu Qiao, and Chen Change Loy. "ESRGAN: Enhanced super-resolution generative adversarial networks." In *Proceedings of the European Conference on Computer Vision (ECCV)* (2018).

Xie, Xingzhi, Zheng Zhong, Wei Zhao, Chao Zheng, Fei Wang, and Jun Liu. "Chest CT for typical 2019-nCoV pneumonia: Relationship to negative RT-PCR testing." *Radiology* (2020): 200343.

4 Application and Progress of Drone Technology in the COVID-19 Pandemic
A Comprehensive Review

Vasundhara Saraf, Lipsita Senapati and Tripti Swarnkar

CONTENTS

- 4.1 Introduction 47
- 4.2 Research Needs..49
- 4.3 A Historical Overview of Drone and UAV Technologies............................50
- 4.4 Different Applications of Drones 54
  - 4.4.1 Agriculture..54
  - 4.4.2 Health Sector 56
  - 4.4.3 Mass Media and Journalism 58
  - 4.4.4 Civil Surveillance 59
  - 4.4.5 Other Applications..60
- 4.5 Opportunity and Challenges...60
- 4.6 Conclusion and Outlook 61
- References...62

4.1 INTRODUCTION

COVID-19 is a transferable disease that has been discovered recently after the Wuhan Chain in China in December 2019 (Li et al., 2020). The disease affected has over 24 million people within a span of 8 months and about 824,162 people have lost their lives to it (Euchi, 2020). Figure 4.1 represents the day-wise growth rate of the cases in highly infected countries of the world. The most common symptoms are fever, tiredness, and dry cough. In some people, symptoms also include throbbing pain, runny and clogged nose, sore throat, nausea, and diarrhea (Rocke et al., 2020). The elderly people and the newborn are more vulnerable to the adverse health effects of this virus. People with low immunity and medical conditions like hypertension, cardiac disease, or diabetes are more likely to be affected by this virus and should seek immediate medical care (Kitchin, 2020). It is a contagious disease and so it spreads between people when they are in close contact through small beads of

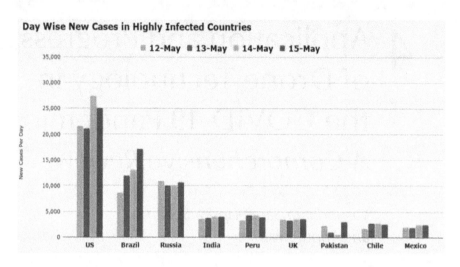

FIGURE 4.1 Graph on COVID-19 growth rate day-wise new cases in highly infected countries of the world in the month of May 2020. Source: www.newsclick.in/.

droplets while talking, coughing, or sneezing (Rocke et al., 2020). The life span of the virus over any surface is up to 72 hrs before it jumps into a host body. Due to the contagious nature of the virus, protective measures include lockdown, handwashing, covering the mouth and nose with a mask, and keeping a 6-ft distance from others (McDougal, 2013; Skorup and Haaland, 2020; Euchi, 2020). This has led to the restriction in mobility of the people, shutdown of all shops, closure of facilities, and also self-isolation. In such a time of pandemic, drone-based COVID-19 healthcare platforms have been explored and improved by scientists across the country. Drones have proven to be a very promising technology in this era of pandemic (Rosser et al., 2018a; Vafea et al., 2020; Hassanalian and Abdelkefi, 2017). Drones are unmanned aerial vehicles, whose most significant usage during the COVID-19 pandemic encourages social distancing (McDougal, 2013).

Drones are playing a very important role in the fight against the coronavirus. They have proven to be a great support for activities taken up by police departments, municipality, and also the health departments of the world. Areas in which drone technology is currently being used include:

a. Lockdown enforcement and surveillance
b. Mass media and broadcasting
c. Monitoring patient's health (recording temperature)
d. Medical and emergency food supplies delivery
e. Mapping and surveying
f. Spraying disinfectants (COVID-19 drones)

Furthermore, the challenges that are faced and are to be covered to make optimum use of the drone technology include processing of collision-free single- and multilayered drone movement strategies and a system which is based on artificial intelligence to collect, analyze, and provide necessary measures (Hassanalian and Abdelkefi, 2017). Proposing a multilayered architecture that collects information from drones and exchanges with fog, cloud, and edge servers for necessary computing, data analytics, and sharing is also one of the greatest challenges (Koubâa and Qureshi, 2018). Moving forward, the challenges include feigning of the drone-based system for COVID-19 operations such as controlling, monitoring, sanitation, social distancing, thermal imaging, medication, data analytics, and generation of statistics for the control room, to implement a real-time drone-based system for sanitization, monitoring, vigilance, face recognition, thermal scanning, etc., in COVID-19 hotspots, and to design and display the statistics of the drone-based smart healthcare system in a control room (Kumar et al., 2020).

This chapter consists of four different sections covering all the important points and sectors in order to conduct proper comprehensive reviews of drone technology and its applications in the COVID-19 pandemic. Section 4.2 is about the research needed to develop drone technology further in order to meet the requirements of society, and also to discuss the problems and solutions that are being worked upon in order to improve the present drone technology. Section 4.3 presents a timeline of the growth of drone technology, from its early use in wars as air-filled balloons to the present-day quadcopters, which are used for delivering groceries and medical supplies and also for monitoring public health. Section 4.4 describes the usage of drones in different sectors. At the end of the chapter, we discuss the opportunities and challenges that are linked with drones and also the possibility of using this technology in order to fight this worldwide pandemic. Moreover, in this chapter, we investigate in detail the research needs and different applications of drones during this pandemic of COVID-19, briefly discussing the growing demand, opportunities, and challenges in the usage of the unmanned aerial vehicles.

4.2 RESEARCH NEEDS

Technology is a key tool for dealing with any problem, so it is no wonder that drone use has soared in recent months, since drones provide undoubted advantages in situations of social confinement and distancing (Lai et al., 2020; Mastaneh and Mouseli, 2020). Drones play an important role by helping authorities and people in different ways and thus preventing the further spread of coronavirus outbreak. Drones are used for fulfilling many purposes such as surveillance, broadcast, disinfectant spraying, medicine and grocery deliveries, and patient monitoring.

We reviewed over 100 recent refereed publications from various electronic databases such as Web of Science, Scopus, Google Scholar, and IEEE explorer to gain a deeper understanding of the research domain. The search terms were "Drone," "UAV devices," "Covid 19," "Virus," "Pandemic," "Social Distancing," "Technology," etc. With the help of a selection filter, papers from the past decade were considered for this review work. A number of papers have analyzed the application of drones in different domains including agriculture, heaths, and hopping. Many civilian applications of drones have been examined by González-Jorge et al. (2017). The application of drones in the delivery of medications and other healthcare items has been reviewed by Judy E. Scott et al. (2019) and Jalel Euchi et al. (Hentati and Fourati, 2020). Aicha Idriss Hentati et al. (Hentati and Fourati, 2020) have presented their views on how drones can be used to improve the communication networks during this pandemic. Many review articles focused on examining the contribution of drones in agriculture (Tsouros et al., 2019; Sharma and Bali, 2018; Messina and Modica, 2020; Boursianis et al., 2020; Hassler and Baysal-Gurel, 2019; Barbedo, 2019; Radoglou et al., 2020) and applications of drones in remote sensing are discussed by Colomina and Molina (2014). Regulations for drones were presented by Stöcker et al. (2017). Uma Shankar Panday et al. (2020) reviewed database solution for cereal crops. A review paper by Brent Skorup and Connor Haaland (2020) sheds light on the application of drones during the COVID-19 pandemic. Potential technological strategies to control the COVID-19 pandemic by Rajvikram Madurai Elavarasan et al. (2020) covers all the fourth-generation technologies that can be made use of in the fight against the coronavirus disease.

The review articles have covered the common use of drones in monitoring citizens, crop growth, delivering medications, providing communication facilities, carrying information and reaching out to the affected areas, and providing assistance to different sectors of the society. There is a lack of focus on the proper mapping, monitoring, and maintenance of the drones. This chapter presents the opportunities for collecting information from citizen science and the Internet of Things (IoT) based on low-cost sensors and drone-based information to Earth Observation (EO) satellite data. The main motive of this chapter is to review cost-effective ways to promote the use of drone technology and improve the working of this technology to benefit all the sectors of the society in such times of the pandemic.

The main contribution of this comprehensive review article is as follows:

1. A historical overview of drone and UAV technologies
2. Application of drones in different sectors
3. Opportunities and challenges
4. Conclusion and outlook

4.3 A HISTORICAL OVERVIEW OF DRONE AND UAV TECHNOLOGIES

An unpiloted aircraft or a spacecraft is called a drone. It is also referred to as an unmanned aerial vehicle or UAV. In technical terms, a drone is an unmanned flying object which is controlled or flies autonomously using software control systems embedded with plans for flight in the system. The embedded software system for flight control of the drones works in conjunction with the Global Positioning System (GPS) and onboard sensors.

A drone consists of motors, propellers, electronic speed controllers (ESC), a flight controller, a radio receiver, and a battery (Ahirwar et al., 2019). The functions of the basic parts of a drone are shown in Figure 4.2 and are discussed in brief in the following sections:

i. Motor: one motor is used for each propeller; the drone motors are rated in kV units. The faster the motor spin is, the more the flight time is, but it also requires a more powerful battery for an increased flight time.
ii. Propellers: three factors mainly depend upon the propellers of the drone: the load they can carry, the maximum speed with which they can fly, and the speed at which they can maneuver. The length of the propellers can be modified according to the requirements.
iii. Electronic speed controller (ESC): the purpose of an electronic speed controller (ESC) is to provide dependent/controlled current to each of the motors attached to the propellers so that they can attain the required spin and speed.
iv. Flight controller (FC): the circuit board which translates the input signals and sends them to the corresponding ESC is called a flight controller (FC).
v. Radio receiver: it receives the controlled signals from the pilot.
vi. Battery: drones use high power density batteries which can be charged easily. Example: lithium polymer batteries.

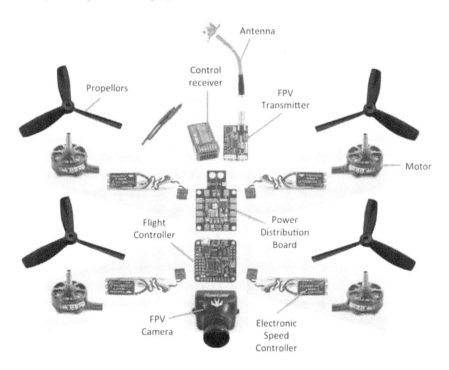

FIGURE 4.2 Different parts and components of a drone or basic anatomy of a drone.
Source: https://fpvdronereviews.com/.

The application of unmanned aerial vehicles can be traced back to the 1800s. With the devising of drones, we entered into a new frontier of war that's risk-free, detached from remote cues, and remote. The earliest history of unmanned aerial vehicle dates back to the bomb-filled air balloons that were first used by Austria in 1839. The invention of the fixed-wing aircraft in 1903 stimulated the production of drones (Giones and Brem, 2017). During World War I, the United States developed drone weapons. Ambrose Sperry was hired by the US army to manufacture and design a column of unmanned air torpedoes that were launched by catapult and aimed at a precise target (McDougal, 2013). These weapons were not deployed until the end of 1918. In World War II, drones emerged as surveillance crafts. They were then improved into remote control unmanned aerial devices. These devices were also used to target the German soldiers who were hiding in the bunkers (McDougal, 2013). Drones were used by the Israel army in 1982 in their war against the Syrian army (Giones and Brem, 2017). The extensive development and manufacturing of military drones began in the early 1980s in the United States after the Cold War. The further development of drones was stalled for a long time until the Central Intelligence Agency (CIA) began covert surveillance missions in Afghanistan in 2000 (McDougal, 2013).

Customs and Border Protection (CBP), a component of Development of Homeland Security (DHS), currently has the largest fleet of drones in operation over US airspace.

> CBP currently deploys nine Predator-B unmanned aerial systems to patrol US borders and provide surveillance for border security and disaster response. Two of these Predators are equipped with sensors for patrolling ports and waterways, thus putting them in a maritime surveillance configuration known as the Guardian. CBP anticipates receiving a 10th Predator UAS in September; it will be the third configured as a Guardian.
>
> **(McDougal, 2013)**

Drone systems have emerged as a favored way of delivering precision strikes with minimal damage to human lives and some control over the event. They have been proven reliable and effective across a wide range of terrains, climates, and battlefield environments.

The quadcopter is the most common configuration of modern-day commercial drones. The Breguet brothers, Jacques and Louis, with the help of a physiologist, Professor Charles Richet, developed one of the very first gyroplanes, a forerunner of the helicopter. The Federal Aviation Administration (FAA) officially issued the first commercial drone permit in 2006; however, the application of drones in the commercial sector was slow to start, as very few people and organizations applied for the permit. The development of the methodology and implementation of drones has gone manifold over the time span of 15 years. In 2013, CEO of Amazon Jeff Bezos announced that the company would consider using drones as a method of delivery. In 2015, the FAA issued 1,000 drone permits. The number of permits issued by the FAA tripled in 2016 when 3,100 permits were issued (McDougal, 2013). The growth graph of the number of permits issued every year continues to grow.

Breakthrough of Drone Technology in COVID-19

Figure 4.3 represents the continuous technological growth of UAV or drone demand as a valuable addition to any industrial facility's efforts, enabling a limitless aerial perspective. The possibilities of drones to create real values are endless, especially in the period of any emergency or pandemic.

The usage of drones during this COVID-19 pandemic opens up new opportunities and generates efficiencies in many sectors discussed in detail in the following sections. Government and private sectors are increasingly using drones to improve and optimize their undoubted advantages in situations of social confinement and distancing. The nearly limitless visibility, data gathering, and analyzing capabilities of the drone make this UAV valuable during this pandemic (Bhatt et al., 2018).

Further advancement in technology saw the equipment of drones with recorders and cameras, thus opening a new possibility of their usage in photography and videography sectors. With the rapid growth in the usage of smartphones, the price of microcontrollers and camera sensors were reduced, which were ideal for use in fixed-wing hobbyist aircraft. Further advancement led to the multi-rotor aircrafts with different speed adjustment features, with each feature thus opening up a new range of possibilities for drones to be used in a number of ways (Vergouw et al., 2016).

Drone technology is growing exponentially, and it is hard to say what more advancements can be expected in the future. The future of drones holds a lot of alterations and modifications.

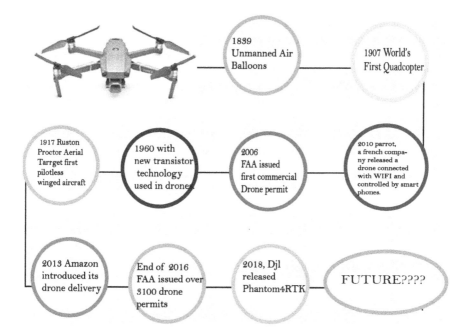

FIGURE 4.3 A timeline showing the historical and technological growth of drone with a limitless future advancement.

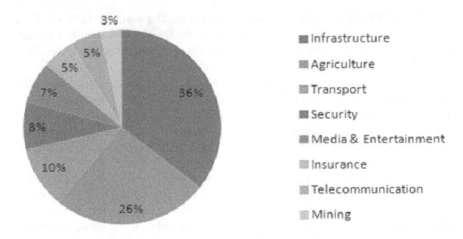

FIGURE 4.4 Represents the significance of drone technology in the commercial world [Hassler, Samuel, and Fulya (2019), Huang, Yanbo, Wesley, Yubin, Wenfu, and Bradley (2019)].

A company named Altair Aerial is manufacturing easy-to-fly drones to get even more people into the drone market with low-cost beginner-friendly drone technology. Companies like EHANG from China want to turn drones into a taxi service for the local public and have built a quadcopter capable of carrying passengers. And Flyability is leading the way to create drones that are able to operate indoors in confined and complex spaces along with people (McDougal, 2013). As seen from Figure 4.3, the history of drones has come a great way from exploding balloons to passenger-carrying unmanned flying taxis.

From the history of drone technology, we can say that drones were initially viewed as a military device. Figure 4.4 represents the growing market of drones in different sectors since their inception as a commercial product. The technological growth of drones has produced a wide range of UAV models of different costs, shapes, and weights, making them favorable for usage across broad applications.

4.4 DIFFERENT APPLICATIONS OF DRONES

As said earlier, drones have their applications in various fields, some of which are reviewed and discussed in this chapter. Automated drones have a wide range of applications, with an undoubted advantage in situations of social confinement and distancing (Sarfo and Karuppannan, 2020; Siriwardhana et al., 2020). Figure 4.5 represents a list of a few applications of drone technology, ensuring safety during the COVID-19 pandemic.

4.4.1 Agriculture

Food security has been a long-standing global issue for the last few centuries. Eradicating hunger and malnutrition from a nation still remains a key challenge for

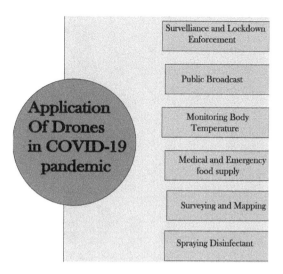

FIGURE 4.5 Applications of drone technology ensuring safety during COVID-19.

experts. The COVID-19 pandemic has placed additional stress on food production, demand, and supply chains. It is assumed that the agricultural needs of the world are going to be increased by 70% in the next few decades (Ahirwar, S., et al., 2019). Annual cereal production will have to rise to 3 billion tons from 2.1 billion tons in the present day and meat production will have to rise to about 200 million tons (Ahirwar, S., et al., 2019). As much as we depend on agriculture to such great extents, we are far from adopting the latest technologies in the agricultural sector. Farmers are always looking for cheap and sustainable methods to regularly monitor their crops. The cameras and sensors in drones are used for this purpose. The infrared sensors of drones can be tuned to detect the health of the crops, enabling farmers to improve the crop health condition locally. Many systems such as power lines, wind turbines, and pipelines can be checked using drones (Ahirwar, S., et al., 2019). Drones are also mounted with sprayer systems for pesticide spraying. The integration of the UAV with the pesticide sprayer is very useful in accurate and specific spraying of chemicals. This requires heavy-lift UAVs for large area spraying (Sarghini and De Vivo, 2017a,b). The efficiency of the spraying system can be increased using the Pulse Width Modulation (PMW) controller (Huang et al., 2009; Zhu et al., 2010).

The first petrol-powered unmanned air vehicle developed in India was the Yamaha RMAX for spraying pesticide in rice fields (Giles and Billing, 2015). An Aerial Automated Pesticide Sprayer (AAPS) was developed based on the Global Positioning System (GPS) coordinate in the lower altitude environment (Vardhan et al., 2014). To overcome the problem of affordability, the cost of the drones was reduced when an android app-controlled drone called the "Freyr" was designed (Spoorthi et al., 2017). Even after so much advancement, there were many challenges to be faced, including the discharge and pressure rate of the liquid which was sprayed, the liquid loss while spraying, uniform spraying density and the size of the

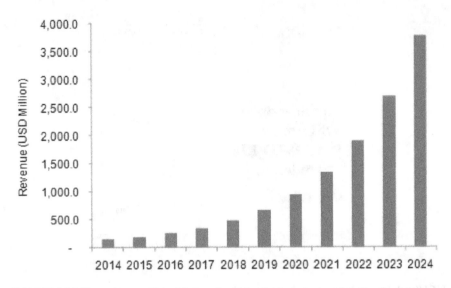

FIGURE 4.6 Emerging market of drones in the agriculture sector globally. Source: www.grandviewresearch.com.

droplets, and the size of the sprayer which was to be mounted (Yallappa et al., 2017). To reduce the wastage of pesticides, an electrostatic sprayer with a hexa-rotor UAV was introduced (Yanliang et al., 2017).

Hyper-spectral and multispectral drones with thermal sensors can identify the parts of the fields that are dry or need improvements. Once the crop starts growing, drones allow calculations of the vegetation index, which describes the relative density and health of the crop, and show the heat signature of the crops (Ahirwar, S., et al., 2019). Drones are a quick means to gather information and navigate through fields after a disaster. During the outbreak of a pandemic like COVID-19, drones come in handy to follow social distancing and thus provide optimum and affordable technologies for agriculture.

The usage of drone technology in the agriculture sector is showing exponential growth globally, as can be seen in Figure 4.6. After the outbreak of the COVID-19 pandemic, the drone-based agriculture market is expected to witness significant growth. Before the pandemic, in the year 2018–2019, a vertical growth of 32% was observed in the usage of drone technology, as can be seen in Figure 4.6. The outbreak of the coronavirus in 2020 increased the drones market from mid-March to mid-April 2020 during the heat of COVID-19 lockdown in the global market, specifically in Asian countries. Figure 4.6 represents that agriculture automation will continue to drive the UAV industry innovation post-pandemic and will possibly revolutionize their process for the better.

4.4.2 HEALTH SECTOR

The most prominent public health measure introduced due to the coronavirus outbreak was social distancing, for which drones can be used for mass service, thus

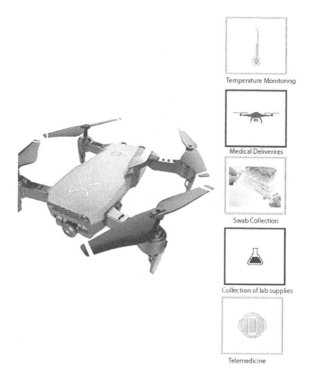

FIGURE 4.7 Usage of drones in the healthcare sector during the COVID-19 pandemic.

reducing person-to-person contact. In the medical sector, the drone is used for medical deliveries, collection of swabs and blood samples for testing, and patient health monitoring. Unmanned aerial vehicles or drones are increasingly explored as a solution to the transportation challenges of medical goods including blood bags, vaccines, diagnostic reports, medicines, and even organs (Wright et al., 2018; Zraick, 2019).

The usage of drones in the healthcare sector is a very attractive and revolutionary idea with many intentional and unintentional consequences (Vafea et al., 2020; Rosser et al., 2018b). The future use of drones in health care is also very provocative. As can be seen in Figure 4.7, this UAV technology is already being used to efficiently provide healthcare services from a distance or while mobile. During outbreaks of deadly communicable diseases like COVID-19, quick and frequent deliveries of vaccines, medications, and other medical emergency–related supplies are not an easy task (Euchi, 2020; Rosser et al., 2018a; Bhatt et al., 2018). The medical services that are required by communities during this pandemic are making medical professionals stressed and they face a lot of challenges every day – drones, being UAVs, can be used to overcome these challenges.

The technological viability of drones to safely transport medical supplies and keep them within the required parameters for clinical viability is proved by concept tests (Amukele et al., 2017; Amukele et al., 2016; Amukele et al., 2015). UAVs have also been identified as a means of adding cost-effectiveness in medical supply chains (Haidari et al., 2016; Euchi, 2020). Any new implementation is affected by a set of factors; in the case of drones being implemented in the medical sector, these factors include the need for transportation within regulations, financial resources availability, and the presence of the human resource at the required destination, and these operational procedures must be developed to work within the existing structure (Dearing, 2018). The impact of drones on the broader health system must be considered before implementation with any kind of intervention (Swanson et al., 2009). These rules for UAVs to be used in the health sector are made using the framework of the World Health Organization (WHO) health system building blocks (World Health Organization, 2007).

Zipline has been operating drones for the Government of Rwanda, delivering up to 3 liters of blood within 30 minutes to health facilities that request it on demand, since 2016. In 2019, several other companies received approval to conduct routine flights. Matternet, a company that has conducted hundreds of test flights in Switzerland, routinely transports laboratory specimens within the North Carolina health system (Eichleay et al., 2019). In addition to these large-scale operations, many other projects deliver medical goods and are implemented worldwide, transporting items like vaccines for children, automatic external defibrillators, snakebite antivenom (McDougal, 2013), tuberculosis sputum samples, sterile mosquitoes, and, presently in some countries, COVID swab samples (Skorup and Haaland, 2020).

The benefits of using drones in healthcare deliveries include (Eichleay et al., 2019; Balasingam, 2017):

- Improved ability to reach patients who require immediate medical attention.
- Ability to transfer medical supplies autonomously.
- Potentially decreases the cost of providing medical care.
- Helps in providing medical care to patients in remote areas.

The challenges associated with using drones in the medical sector are (Vergouw et al., 2016; Kitchin, 2020):

- The integrity of the specimen should be maintained during payload capacity
  - Temperature
  - Special packaging
- Security for controlled substances
- Battery life
- Approval from the authorities
- The demand of the customers

4.4.3 Mass Media and Journalism

The COVID-19 pandemic has newsrooms across the country and all over the world scrambling to deliver vital information to their audiences. News publications and reporters have taken on new skills at a time where the public is intently tuned in to each breaking news about the virus and the disease it causes. The idea of mapping and obtaining aerial images from a considerable distance is very alluring for journalists and photographers.

The application of drones in journalism includes capturing and obtaining images when the circumstances are not suitable for reports to be sent to certain areas or localities, such as armed conflicts, spread of a contagious disease, and other adverse conditions. Also when the channels have to show an image of an area which is practically inaccessible by humans, they use drones. The drones help in considerably reducing the cost of obtaining better images, with perspective, speed, mobility, and greater safety.

The advantages of drones as a tool for news coverage are:

- They contribute a visual value to news coverage; for example, in a strike or a fight, the actual number of people involved could be recorded with the help of aerial footage.
- Covering sports events from all angles.
- Real-time broadcasting.
- Monitoring activities of certain organizations without being noticed.
- Footage of areas which are practically inaccessible.
- And in the present case, helping people to minimalize person-to-person contact and recording the footage of the affected areas, thus communicating the messages of the affected people to the outside world.

4.4.4 CIVIL SURVEILLANCE

The government and civil service organizations have been constantly working toward a maneuver and an approach to fight and mitigate the effect of the COVID-19 pandemic. To augment and supplement the traditional measures like personal hygiene, sanitation, social distancing and self-isolation, using masks, avoiding direct contact and also visiting public places, and enforced quarantining and testing regimes, existing and new digital technologies are being harnessed. These technologies are supported by the arguments that they will help optimize population control, by improving the real-time mass monitoring at the individual and aggregate level (Hossain et al., 2020). We reviewed the different applications of drones in civil surveillance in different countries. The primary purposes on which the countries focused in order to get a technological aid were:

- Recording the data of how many people live in which area and enforcing lockdown/quarantine.
- Approved movement for those with permission and enforcing home isolation for the infected people.
- Contact tracing.
- Tracking the pattern of the spread of the virus.
- Keeping a check on people adhering to proper social distancing measures.
- Circulating information and announcements in inaccessible areas.
- Symptoms tracking.

Countries like Italy have been using thermal cameras to monitor the body temperature of people in public places and also to police the breaking of lockdown. The adoption of drone technology at the civilian level is in its nascent stage. Drones have been used for surveillance, announcements, monitoring, and guiding local law enforcement during the current pandemic (Lee et al., 2018). Infrared cameras are mounted to record the body temperature. China implemented drone surveillance and broadcast special instructions for safeguarding against the coronavirus. Further, these drones are equipped with QR code placards that could be scanned to register health information (Boulos and Geraghty, 2020). Kazakhstan used a drone with infrared sensors for patrolling and monitoring illegal border movements to slow the spread of coronavirus infections. These drones are provided by the Terra Drone group company, KazUAV (Lee et al., 2018). In India, Gujarat and Kerala police deployed 200 and 500 drones, respectively. Further, police in highly populated cities like Delhi, Mumbai, and Kolkata also deployed drones for mass surveillance and monitoring (Lee et al., 2018). Several police forces in the UK used drones to monitor people breaking rules by visiting national parks. The drone footage was used to highlight correct and incorrect behaviors (Lee et al., 2018).

4.4.5 Other Applications

One of the imperatives imposed on societies affected by the COVID-19 is improved hygiene. Drones have emerged as an effective tool that can sanitize large spaces and help lower the risk of infection from coronavirus for humans. More than 900 km^2 area has been disinfected in China (Euchi, 2020). In Korea, drones sprayed over an area of 10,000 m^2 in around 10 minutes.

Drones were also used for delivering daily need items to the restricted and affected areas during this pandemic (Kellermann et al., 2020). This helped reduce the spread of the virus and also ensured an optimal supply of food, water, and other necessities to the people who were affected (van Veelen et al., 2020).

4.5 OPPORTUNITY AND CHALLENGES

With artificial intelligence (AI) and the Internet of Things (IoT) gaining popularity in this era, it won't be hard to believe that the future of the drone industry is going to be more of these autonomous flying vehicles, which would be the end result of amalgamation of AI and IoT (Koubâa and Qureshi, 2018; Lee et al., 2018). These ultra-smart flying machines would use computer vision, equipped with deep learning and machine learning algorithms. These would create havoc in the drone kingdom as it will be able to perform obstacle detection, object tracking, self-navigation, and collision avoidance by virtue of computer vision (Bartak and Vykovský, 2015).

When it comes to the futuristic scope of unmanned aerial vehicles like drones, the sky is the limit. Drones by design are meant to be researchers' eye candy which they could use to reach the skies and capture visual information and subsequently transform those into data points and further use them in several aspects ranging from monitoring to providing help. It's going to help humankind in various ways;

search and rescue missions being the first and foremost domain where these drones can save lives where humans are seen on the edge (Singla, 2020). Apart from this, in no time drones will be seen as the knight in shining armor in the entire aftermath phase of any man-made or natural disasters. These unmanned miniature vehicles would be seen identifying, mapping, monitoring, and accessing the manufactural damage caused in buildings, bridges, roads, monuments, tunnels, etc., which sometimes is practically impossible for humans to perform. The most important task of post-disaster management is the distribution of relief to the disaster victims and delivering emergency infrastructures and supplies (Chowdhury et al., 2017). This important task could be easily handled by supersonic drones in the future. Medical drones are expected to play a major role in bridging the ever-long gaps between the advanced healthcare systems of metropolitan cities and the poor healthcare units of distant rural areas via telemedicine assignments. This will ultimately lead to a huge leap in the medical sector. For uncountable reasons, drones can be visualized as a lifesaving technological advancement that has vast potential to be well utilized for humankind in the future. The future of drones won't only be confined to the sky; rather it will work wonders in oceans and seas as well, with advancement in deep sea drones, aiming at exploring new archaeological/geological/biological features (Toyama and Nishizawa, 2019).

A coin has two sides. No doubt drones have emerged as lifelines in this era, but they also come with a huge factor of risk: risk of criminal activities witnessing drones equipped with weapons, risk of violating privacy policy, risk of terrorism activities, etc. (Hassanalian and Abdelkefi, 2017). So, with marvelous opportunities, it also comes with challenges to take care of; right from the day drones came into our lives, they have raised questions on safety, security, privacy, and regulation, which are undoubtedly the areas we need to work on (Rao et al., 2016). Apart from that, limitations in payload capacity and flight endurance are often challenging. Thinking of adding extra sources of mechanisms for the enhancement of capacity leads to an exponential increase in the cost of the manufacturing and a decrease in flight time. In addition to this, the entire process of executing these drones for their respective uses may face improper air traffic management issues to tackle.

4.6 CONCLUSION AND OUTLOOK

Drones, which were once upon a time more likely to be used for one or two generic purposes, in the course of time have established their integral position in various arenas, and their role in this pandemic scenario is highly pivotal when everyone is stuck indoors for good. This review attempts to provide a comprehensive understanding of the research needs and different applications of drones during this pandemic of COVID-19. The growing demand, opportunities, and challenges in the usage of unmanned aerial vehicles in different application areas to help reduce the impact of the COVID-19 outbreak are briefly discussed. In addition to the direct health implications, this chapter throws light on the socioeconomic impact of the technology globally.

After knowing that COVID-19 won't be the last pandemic, as stated clearly by the World Health Organization in September 2020, the world needs to be more prepared than it ever was with these emerging technological wonders in order to combat pandemics and outbreaks like never before. Undoubtedly in a scenario like this, the present and future of drones look bright.

REFERENCES

Ahirwar, S., et al. "Application of drone in agriculture." *International Journal of Current Microbiology and Applied Sciences* 8, no. 1 (2019): 2500–2505.

Amukele, Timothy K., Lori J. Sokoll, Daniel Pepper, Dana P. Howard, and Jeff Street. "Can unmanned aerial systems (drones) be used for the routine transport of chemistry, hematology, and coagulation laboratory specimens?" *PloS one* 10, no. 7 (2015): e0134020.

Amukele, Timothy K., Jeff Street, Karen Carroll, Heather Miller, and Sean X. Zhang. "Drone transport of microbes in blood and sputum laboratory specimens." *Journal of Clinical Microbiology* 54, no. 10 (2016): 2622–2625.

Amukele, Timothy K., James Hernandez, Christine L.H. Snozek, Ryan G. Wyatt, Matthew Douglas, Richard Amini, and Jeff Street. "Drone transport of chemistry and hematology samples over long distances." *American Journal of Clinical Pathology* 148, no. 5 (2017): 427–435.

Balasingam, Manohari. "Drones in medicine—the rise of the machines." *International Journal of Clinical Practice* 71, no. 9 (2017): e12989.

Barbedo, Jayme Garcia Arnal. "A review on the use of unmanned aerial vehicles and imaging sensors for monitoring and assessing plant stresses." *Drones* 3, no. 2 (2019): 40.

Bartak, Roman, and Adam Vykovský. "Any object tracking and following by a flying drone." In *2015 Fourteenth Mexican International Conference on Artificial Intelligence (MICAI)*, pp. 35–41. IEEE, 2015.

Bhatt, Kunj, Ali Pourmand, and Neal Sikka. "Targeted applications of unmanned aerial vehicles (drones) in telemedicine." *Telemedicine and e-Health* 24, no. 11 (2018): 833–838.

Boulos, Maged N. Kamel, and Estella M. Geraghty. "Geographical tracking and mapping of coronavirus disease COVID-19/severe acute respiratory syndrome coronavirus 2 (SARS-CoV-2) epidemic and associated events around the world: how 21st century GIS technologies are supporting the global fight against outbreaks and epidemics." (2020): 1–12.

Boursianis, Achilles D., Maria S. Papadopoulou, Panagiotis Diamantoulakis, Aglaia Liopa-Tsakalidi, Pantelis Barouchas, George Salahas, George Karagiannidis, Shaohua Wan, and Sotirios K. Goudos. "Internet of Things (IoT) and agricultural Unmanned Aerial Vehicles (UAVs) in smart farming: a comprehensive review." *Internet of Things* (2020): 100187.

Chowdhury, Sudipta, Adindu Emelogu, Mohammad Marufuzzaman, Sarah G. Nurre, and Linkan Bian. "Drones for disaster response and relief operations: a continuous approximation model." *International Journal of Production Economics* 188 (2017): 167–184.

Colomina, Ismael, and Pere Molina. "Unmanned aerial systems for photogrammetry and remote sensing: a review." *ISPRS Journal of Photogrammetry and Remote Sensing* 92 (2014): 79–97.

Dearing, James W. "Organizational readiness tools for global health intervention: a review." *Frontiers in Public Health* 6 (2018): 56.

Eichleay, Margaret, Emily Evens, Kayla Stankevitz, and Caleb Parker. "Using the unmanned aerial vehicle delivery decision tool to consider transporting medical supplies via drone." *Global Health Science and Practice* 7, no. 4 (2019): 500–506.

Elavarasan, Rajvikram Madurai, and Rishi Pugazhendhi. "Restructured society and environment: a review on potential technological strategies to control the COVID-19 pandemic." *Science of The Total Environment* 725(2020): 138858.

Euchi, Jalel. "Do drones have a realistic place in a pandemic fight for delivering medical supplies in healthcare systems problems." *Chinese Journal of Aeronautics* (2020).

Giles, D., and R. Billing. "Deployment and performance of a UAV for crop spraying." *Chemical Engineering Transactions* 44 (2015): 307–312.

Giones, Ferran, and Alexander Brem. "From toys to tools: the co-evolution of technological and entrepreneurial developments in the drone industry." *Business Horizons* 60, no. 6 (2017): 875–884.

González-Jorge, Higinio, Joaquin Martínez-Sánchez, and Martín Bueno. "Unmanned aerial systems for civil applications: a review." *Drones* 1, no. 1 (2017): 2.

Haidari, Leila A., Shawn T. Brown, Marie Ferguson, Emily Bancroft, Marie Spiker, Allen Wilcox, Ramya Ambikapathi, Vidya Sampath, Diana L. Connor, and Bruce Y. Lee. "The economic and operational value of using drones to transport vaccines." *Vaccine* 34, no. 34 (2016): 4062–4067.

Hassanalian, Mostafa, and Abdessattar Abdelkefi. "Classifications, applications, and design challenges of drones: a review." *Progress in Aerospace Sciences* 91 (2017): 99–131.

Hassler, Samuel C., and Fulya Baysal-Gurel. "Unmanned aircraft system (UAS) technology and applications in agriculture." *Agronomy* 9, no. 10 (2019): 618.

Hentati, Aicha Idriss, and Lamia Chaari Fourati. "Comprehensive survey of UAVs communication networks." *Computer Standards & Interfaces* 72(2020): 103451.

Hossain, M. Shamim, Ghulam Muhammad, and Nadra Guizani. "Explainable AI and mass surveillance system-based healthcare framework to combat COVID-19 like pandemics." *IEEE Network* 34, no. 4 (2020): 126–132.

Huang, Yanbo, Wesley C. Hoffmann, Yubin Lan, Wenfu Wu, and Bradley K. Fritz. "Development of a spray system for an unmanned aerial vehicle platform." *Applied Engineering in Agriculture* 25, no. 6 (2009): 803–809.

Kellermann, Robin, Tobias Biehle, and Liliann Fischer. "Drones for parcel and passenger transportation: a literature review." *Transportation Research Interdisciplinary Perspectives* 4 (2020): 100088.

Kitchin, Rob. "Civil liberties or public health, or civil liberties and public health? Using surveillance technologies to tackle the spread of COVID-19." *Space and Polity* 24(2020): 1–20.

Koubâa, Anis, and Basit Qureshi. "Dronetrack: cloud-based real-time object tracking using unmanned aerial vehicles over the internet." *IEEE Access* 6 (2018): 13810–13824.

Kumar, Adarsh, Kriti Sharma, Harvinder Singh, Sagar Gupta Naugriya, Sukhpal Singh Gill, and Rajkumar Buyya. "A drone-based networked system and methods for combating coronavirus disease (COVID-19) pandemic." *Future Generation Computer Systems* 115 (2020): 1–19.

Lai, Ka Yan, Chris Webster, Sarika Kumari, and Chinmoy Sarkar. "The nature of cities and the COVID-19 pandemic." *Current Opinion in Environmental Sustainability* 46 (2020): 27–31.

Lee, Dongkyu Rroyr, Woong Gyu La, and Hwangnam Kim. "Drone detection and identification system using artificial intelligence." In *9th International Conference on Information and Communication Technology Convergence, ICTC 2018*, pp. 1131–1133. Institute of Electrical and Electronics Engineers Inc., 2018.

Li, Lifang, Qingpeng Zhang, Xiao Wang, Jun Zhang, Tao Wang, Tian-Lu Gao, Wei Duan, Kelvin Kam-fai Tsoi, and Fei-Yue Wang. "Characterizing the propagation of situational information in social media during covid-19 epidemic: a case study on weibo." *IEEE Transactions on Computational Social Systems* 7, no. 2 (2020): 556–562.

Mastaneh, Zahra, and Ali Mouseli. "Technology and its solutions in the era of COVID-19 crisis: a review of literature." *Evidence Based Health Policy, Management and Economics* 4, no. 2 (2020): 138–149.

McDougal, Cameron. "From the battlefield to domestic airspace: an analysis of the evolving roles and expectations of drone technology." *PublicINReview* 1, no. 2 (2013): 92–102.

Messina, Gaetano, and Giuseppe Modica. "Applications of UAV thermal imagery in precision agriculture: state of the art and future research outlook." *Remote Sensing* 12, no. 9 (2020): 1491.

Panday, Uma Shankar, Arun Kumar Pratihast, Jagannath Aryal, and Rijan Bhakta Kayastha. "A review on drone-based data solutions for cereal crops." *Drones* 4, no. 3 (2020): 41.

Radoglou-Grammatikis, Panagiotis, Panagiotis Sarigiannidis, Thomas Lagkas, and Ioannis Moscholios. "A compilation of UAV applications for precision agriculture." *Computer Networks* 172 (2020): 107148.

Rao, Bharat, Ashwin Goutham Gopi, and Romana Maione. "The societal impact of commercial drones." *Technology in Society* 45 (2016): 83–90.

Rocke, John, Claire Hopkins, Carl Philpott, and Nirmal Kumar. "Is loss of sense of smell a diagnostic marker in COVID-19: a systematic review and meta-analysis." *Clinical Otolaryngology* 45, no. 6 (2020): 914–922.

Rosser Jr, James Butch, Brett C. Parker, and Vudatha Vignesh. "Medical applications of drones for disaster relief: a review of the literature." *Surgical Technology International* 33 (2018a): 17–22.

Rosser Jr, James C., Vudatha Vignesh, Brent A. Terwilliger, and Brett C. Parker. "Surgical and medical applications of drones: a comprehensive review." *JSLS: Journal of the Society of Laparoendoscopic Surgeons* 22, no. 3 (2018b).

Sarfo, Anthony Kwabena, and Shankar Karuppannan. "Application of geospatial technologies in the covid-19 fight of Ghana." *Transactions of the Indian National Academy of Engineering* 5, no. 2 (2020): 193–204.

Sarghini, Fabrizio, and Angela De Vivo. "Analysis of preliminary design requirements of a heavy lift multirotor drone for agricultural use." *Chemical Engineering Transactions* 58 (2017a): 625–630.

Sarghini, Fabrizio, and Angela De Vivo. "Interference analysis of an heavy lift multirotor drone flow field and transported spraying system." *Chemical Engineering Transactions* 58 (2017b): 631–636.

Scott, Judy E., and Carlton H. Scott. "Models for drone delivery of medications and other healthcare items." In *Unmanned Aerial Vehicles: Breakthroughs in Research and Practice*, pp. 376–392. IGI Global, 2019.

Sharma, Lakesh K., and Sukhwinder K. Bali. "A review of methods to improve nitrogen use efficiency in agriculture." *Sustainability* 10, no. 1 (2018): 51.

Singla, Parveen. "Drone technology-game changer to fight against COVID-19." *Tathapi With ISSN 2320-0693 is an UGC CARE Journal* 19, no. 6 (2020): 78–80.

Siriwardhana, Yushan, Chamitha De Alwis, Gurkan Gur, Mika Ylianttila, and Madhusanka Liyanage. "The fight against COVID-19 pandemic with 5G technologies." *IEEE Engineering Management Review* 48 (2020): 72–84.

Skorup, Brent, and Connor Haaland. "How drones can help fight the Coronavirus." *Mercatus Center Research Paper Series, Special Edition Policy Brief (2020)* (2020).

Spoorthi, S., B. Shadaksharappa, S. Suraj, and V.K. Manasa. "Freyr drone: pesticide/fertilizers spraying drone-an agricultural approach." In *2017 2nd International Conference on Computing and Communications Technologies (ICCCT)*, pp. 252–255. IEEE, 2017.

Stöcker, Claudia, Rohan Bennett, Francesco Nex, Markus Gerke, and Jaap Zevenbergen. "Review of the current state of UAV regulations." *Remote Sensing* 9, no. 5 (2017): 459.

Swanson, R. Chad, Henry Mosley, David Sanders, David Egilman, Jan De Maeseneer, Mushtaque Chowdhury, Claudio F. Lanata, Kirk Dearden, and Malcolm Bryant. "Call for global health-systems impact assessments." *The Lancet* 374, no. 9688 (2009): 433–435.

Toyama, Shigeki, and Uichi Nishizawa. "Deep-sea drone with spherical ultrasonic motors." *International Journal of Modeling and Optimization* 9, no. 6 (2019): 348–351.

Tsouros, Dimosthenis C., Stamatia Bibi, and Panagiotis G. Sarigiannidis. "A review on UAV-based applications for precision agriculture." *Information* 10, no. 11 (2019): 349.

Vafea, Maria Tsikala, Eleftheria Atalla, Joanna Georgakas, Fadi Shehadeh, Evangelia K. Mylona, Markos Kalligeros, and Eleftherios Mylonakis. "Emerging technologies for use in the study, diagnosis, and treatment of patients with COVID-19." *Cellular and Molecular Bioengineering* 13, no. 4 (2020): 249–257.

van Veelen, Michiel J., Marc Kaufmann, Hermann Brugger, and Giacomo Strapazzon. "Drone delivery of AED's and personal protective equipment in the era of SARS-CoV-2." *Resuscitation* 152 (2020): 1–2.

Vardhan, P.D.P.R. Harsh, S. Dheepak, P.T. Aditya, and Sanjivi Arul. "Development of automated aerial pesticide sprayer." *International Journal of Engineering Science and Research Technology* 3, no. 4 (2014): 458–462.

Vergouw, Bas, Huub Nagel, Geert Bondt, and Bart Custers. "Drone technology: types, payloads, applications, frequency spectrum issues and future developments." In *The Future of Drone use*, pp. 21–45. TMC Asser Press, The Hague, 2016.

World Health Organization. *Everybody's Business: Strengthening Health Systems to Improve Health Outcomes: WHO's Framework for Action*. WHO Health Systems and Services (HSS), Geneva, 2007 [cited 2019 Jan 24].

Wright, Chris, Sidharth Rupani, Kameko Nichols, Yasmin Chandani, and M. Machagge. "What should you deliver by unmanned aerial systems?" *White Paper for JSI Research & Training Institute, Inc., and Llamasoft* (2018).

Yallappa, D., M. Veerangouda, Devanand Maski, Vijayakumar Palled, and M. Bheemanna. "Development and evaluation of drone mounted sprayer for pesticide applications to crops." In *2017 IEEE Global Humanitarian Technology Conference (GHTC)*, pp. 1–7. IEEE, 2017.

Yanliang, Zhang, Lian Qi, and Zhang Wei. "Design and test of a six-rotor unmanned aerial vehicle (UAV) electrostatic spraying system for crop protection." *International Journal of Agricultural and Biological Engineering* 10, no. 6 (2017): 68–76.

Zhu, Hang, Yubin Lan, Wenfu Wu, W. Clint Hoffmann, Yanbo Huang, Xinyu Xue, Jian Liang, and Brad Fritz. "Development of a PWM precision spraying controller for unmanned aerial vehicles." *Journal of Bionic Engineering* 7, no. 3 (2010): 276–283.

Zraick, K. "Like 'Uber for Organs': drone delivers kidney to Maryland woman." *New York Times*. April 30 (2019).

5 Smart War on COVID-19 and Global Pandemics
Integrated AI and Blockchain Ecosystem

*Anil D. Pathak, Debasis Saran,
Sibani Mishra, Madapathi Hitesh,
Sivaiah Bathula, and Kisor K. Sahu*

CONTENTS

5.1	Introduction: COVID-19 in the Era of Modern Technology	68
5.2	Tracking and Forecasting the COVID-19 Outbreak in Real Time: Role of AI	70
5.3	Data Authentication: A Case for Blockchain Integration for Diverse AI Systems	71
5.4	Role of AI in Diagnosis, Detection, and Identification of COVID-19 Patients	73
5.5	Accelerating Drug Discovery for COVID-19: The Big Role of AI	75
5.6	Securing Different Supply Chains at Time of Crisis	77
	5.6.1 Medical Supply Chain	77
	5.6.2 Food Supply Chain	77
	5.6.3 Relocation of Medical, Law Enforcement, and Other Critical Service Professionals	79
5.7	Impact of COVID-19 on World Economy and Role of AI and Blockchain	79
5.8	Other Smart Interventions	81
	5.8.1 Facial Recognition and Fever Detector AI	81
	5.8.2 AI-Enabled Drones and Robots	81
	5.8.3 Virtual Healthcare Assistants (Chatbots)	82
	5.8.4 Managing Finance/Donation (Blockchain)	83
5.9	Outlook/Perspective	83
	5.9.1 Trust: The Missing Piece in Today's Societal and Technological Architecture	83
	5.9.2 Integrated AI and Blockchain Ecosystem: Smart Defense against Global Pandemics	84

5.10 Future Research Directions 84
   5.10.1 The Big Canvas of Integration of Emerging
       Technological Paradigms... 84
   5.10.2 Development of Integrated Frameworks 85
5.11 Conclusion 86
References.. 87

5.1   INTRODUCTION: COVID-19 IN THE ERA OF MODERN TECHNOLOGY

The ongoing global pandemic caused by the virus named "SARS-CoV-2" or popularly known as "Coronavirus" (though this is an incorrect nomenclature since it indicates an entire family of viruses, of which SARS-CoV-2 is only one member) has created an unprecedented global health, financial, societal, and humanitarian crisis unforeseen in the post-internet modern era. The unprecedented rapid development in many diverse areas like medical sciences, drug and pharmaceutical research, information and technology, machine learning, and artificial intelligence (AI) had provided a perfect excuse for a false sense of security and invincibility of the modern-day civilization. It took a concerted effort by a creature that is not even a living organism (the virus is considered an intermediate state between living and dead) to shatter the myth. It has created mayhem throughout the world and even the entire human civilization seems ill-prepared to counter the threat. Therefore, it is imperative to reorient our technological paradigm that is well-prepared to deal with these kinds of situations. This chapter serves the purpose of surveying research articles, articulating emerging trends that are likely to shape the future technological paradigm and keep us prepared to face such eventualities. However, before we start this survey in the next section, let us review the prevailing technological paradigm and identify its critical vulnerabilities.

One of the marvels of modern-day civilization is the internet, which liberated and "democratized" information. While the internet has provided near-universal access (at least in principle), it is associated with lots of challenges. For example, with ready access to information also comes the "disinformation" and at a time of accelerated online fraudulent and even criminal activities. The element of "trust" was the missing piece of the puzzle in this newfound online territory. In a traditional framework, trust is enforced by different regulatory/legal/enforcement agencies in different countries. Disparate legal and enforcement frameworks between countries leads to tremendous heterogeneities across countries, and in general, they are at odds with each other, sometimes even at conflict. However, when the internet emerged toward the end of the previous century, it had little regard for these geographical boundaries, which led to great difficulties in enforcing the jurisdiction of those regulatory agencies over the internet. Therefore, there were desperate needs for development of a technological solution for enforcing "trust" between two or more animate or inanimate agencies in this digital space. In this backdrop, Satoshi Nakamoto (a pseudonym, original name still unknown) presented a whitepaper outlining the first blueprint of "Blockchain," which is the technical manifestation of building trust in the digital space without the intervention of any central agencies by adopting distributed computing protocol. Therefore, it has the requisite potential to establish trust in the digital space between any two or more agents and make them amenable to perform business/transaction/exchange without bothering to dig very deep to establish their credibility. Blockchain is a record of events which can be shared and accessed by a network of computers. It is managed by a peer-to-peer network that adheres to a protocol for inter-node communication and inclusion of new blocks. They are considered secure by design, because the data in any given block cannot be altered retroactively without the alteration of all subsequent blocks (Zīle and Strazdiņa 2018).

Another marvel of modern-day technology is the omnipresent role of machine learning (ML) and AI that has dramatically altered the face of humanity. Technically, ML is a subset of AI, but in recent times, ML has grown so much, at times, we will make separate mention of it in order to impart clarity. Today's digital world is creating data at an unthinkable pace. Harnessing the inherent intelligence hidden deep within this data can create huge potential for new opportunities. It is making our life comfortable, reducing workload, helping to perfect skills, helping humankind in our relentless search for new materials, more effective drugs, just to name a few. It is standing shoulder to shoulder with its human counterpart to make faster discoveries, product development, and marketing. It has brought about unthinkable efficiencies even in many traditional systems. With every passing day, as the research details grow, the amount of data has ascended gigantically, surpassing the ability of human intelligence to handle such enormous data. In this scenario, AI can be useful as it is proficient at identifying patterns from big data. However, AI thrives on a large amount of authentic data. Though the internet is the perfect source for a large amount of data, as we already discussed, there are serious issues about the sanctity of many of these data. We cannot afford to build our critical first line of defense using such a vulnerable source of data. Therefore, it makes the perfect case for marriage between the AI and blockchain for critical and smart infrastructure. A rough sketch of such a type of integrated infrastructure is depicted in Figure 5.1. Though any large-scale deployment of such a futuristic framework to fight pandemics has not happened to the best of our knowledge, it is important to appreciate that all the individual components of this framework are either already developed or at different stages of development.

This chapter focuses on different relevant themes related to emerging technologies in tackling the COVID-19 pandemic and is organized as follows. First, in Section 5.2 of this chapter, the importance of artificial intelligence in tracking and forecasting the COVID-19 outbreak in real time is highlighted. Then, the role of decentralized ecosystems in providing a secure environment for data authentication is discussed in Section 5.3 of this chapter. In Section 5.4, the significance of artificial intelligence in the diagnosis and detection of COVID-19 patients has been explored. Section 5.5 focuses on COVID-19 drug discovery using artificial intelligence techniques. The role of the blockchain platform in securing trust among various supply chains such as food and medical supply chains has been discussed in Section 5.6 of this chapter. Section 5.7 of this chapter focuses on the impact of COVID-19 on the world economy. In Section 5.8, some of the recent developments in artificial intelligence

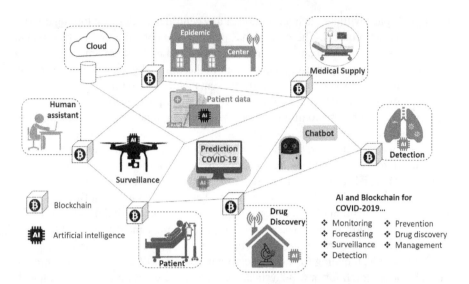

FIGURE 5.1 A schematic showing the highly integrated network topology of AI and blockchain that might create a first line of defense against COVID-19 or other pandemics in the future.

have been considered, and finally, in Section 5.9, the future outlook of this study has been discussed with concluding remarks.

5.2 TRACKING AND FORECASTING THE COVID-19 OUTBREAK IN REAL TIME: ROLE OF AI

COVID-19 is a highly transmittable viral infection from human to human and has a relatively long incubation period (Q. Li et al. 2020; Zou et al. 2020). During this long incubation period, infected persons might show mild or no symptoms. This is a major challenge from a disease containment point of view, and the role of a forecasting model becomes hugely important, since it will help in identifying potentially high-risk zones and super-spreaders so that preventive measures can be taken. Indeed, forecasting models for growth and geographic distribution of the outbreak, dynamics of transmission, targeting resources, and evaluating the impact of intervention strategies and the evolution of hot spots in the outbreak are a very important aspect. All of these may be based upon ML and AI, which might turn out to be very useful tools for understanding the epidemiology of the viral infection and predict its potential impact on public health in local and global communities (Abhari et al. 2020; Hu et al. 2020).

These models can help governments or regulatory bodies to understand how COVID-19 outbreaks evolve, and control measures that need to be taken/altered for controlling the outbreak including source containment, case management, policy directives for schools and organizations, contact tracing, protocols for closing borders, suspending/limiting community services, infection control at healthcare facilities, and community containment (Perrella et al. 2020; Zeng et al. 2020). Globally, serious efforts are going on in the direction of tracking the COVID-19 outbreak with the assistance of AI, and some of them are sampled in the following sections.

Abhari et al. (2020) deployed a growth prediction and containment of COVID-19 in Switzerland using a previously developed AI model (EnerPOL) and big data simulation platform and predicted there will be 720, 73,300, and 83,300 deaths, recoveries, and infections, respectively, between 22 February and 11 April 2020. This model also reckons for pre-intervention as well as post-intervention transmission rate, incubation period, and others.

It is quite reasonable to expect that, at the initial stage of an epidemic/pandemic outbreak, there will be a paucity of useful and reliable data, which makes forecasting a very challenging proposition. Therefore, Fong et al. (2020) proposed a methodology named Group of Optimized and Multisource Selection, abbreviated as GROOMS, to forecast the COVID-19 outbreak from a small dataset. This methodology is an ensemble of five different types of forecasting methods including classical time series forecasting and self-evolving polynomial neural networks (PNN). This study observed that PNN has relatively better performance and lowest error rate and further extended to PNN+cf with corrective feedback in the optimization process (Fong et al. 2020).

Zixin Hu et al. (2020) deployed a modified stacked auto-encoder AI model for real-time forecasting of COVID-19 and also estimated the size, lengths, and ending time of the pandemic. The data spanned from 11 January to 27 February 2020 and forecasted confirmed cases of COVID-19 from 20 January to 20 April 2020 in China. Further, dynamic patterns of the transmission of virus across the provinces/cities were clustered, and the model also predicted that COVID-19 epidemics would be over by the middle of April (Hu et al. 2020), though retrospectively, it turned out to be an inaccurate proposition for the larger part of the globe. Therefore, it is warranted to be aware of the possible pitfalls of model development and reliance on their predictions before they have been validated by extensive research.

AI can be applied to track the coronavirus outbreak by requisite information from social media, websites, news reports, and other sources of data related to COVID-19 symptoms such as fever, breathing, and coughing as indicators, and use this information to predict where the disease is most likely to spread. For example, a company called BlueDot is using foreign-language news reports, animal and plant disease networks, and other official data sources for their AI model to predict, alert, and give advance warning to its clients about the next hotspots (Niller 2020). Thus, it can be argued that AI can cause a paradigm shift to the current COVID-19 or future global pandemic outbreaks by tracking and predicting the outbreak and curbing the spread.

5.3 DATA AUTHENTICATION: A CASE FOR BLOCKCHAIN INTEGRATION FOR DIVERSE AI SYSTEMS

Currently, there is no vaccine available with proven clinical efficacy against COVID-19 (Cennimo 2020). Several clinical trials are ongoing and the World Health Organization (WHO) is playing a central role in reviewing the evidence generated by these trials. However, COVID-19 cases are expected to rise substantially in the coming days, but the supply of protective equipment, testing kits, healthcare staffs, etc., are limited (Livingston et al. 2020) and may lead to overwhelming of healthcare systems in many parts of the world as happened in Italy. To address the ongoing COVID-19 outbreak and accelerate the research with limited resources, we need open science, data sharing, and collaboration between researchers, academia, governments, official organizations, civil society, and the private sector to better monitor, understand, and accelerate COVID-19 research to mitigate this pandemic (United Nations SDSN Report 2020; White 2020). Even crowdsourcing can be a very effective, economical, and robust tool. Smith et al. (2015) discussed various forms of crowdsourcing and their advantages and studied the impact of crowdsourcing on economic resilience with the help of a case study.

One of the main problems with such open science and data sharing strategies is the availability of safe, secure, scalable, and verified data which will require collective efforts of clinicians, scientists, and researchers. Nevertheless, researchers are using blockchain technology in an attempt to mitigate some of these issues by distributed consensus algorithms which possess the following features: decentralized, transparent, immutable data along with autonomy, open source, and anonymity (Liang et al. 2016).

Shen et al. (2019) studied efficient management of patient medical records and proposed a mechanism for sharing medical records using the MedChain model. The model was developed on a decentralized platform, connecting various healthcare stakeholders utilizing the blockchain and peer-to-peer networks. It employed a session-based data sharing scheme which included data generation, session management, and key management. The data generation process consisted of collecting patient data through medical devices and integrating it into blockchain and directory services. During the session management process, a patient can concede access to their medical data to a requester through a session following different protocols. Key management applications in basic terms provided each participant with a software key case for storing received cryptographic keys. They have evaluated the performance and accuracy of this entire system based on two critical parameters, i.e., security and efficiency. Results indicated that the MedChain scheme was highly flexible, efficient, and secure compared to other existing models, without compromising on potential industrial scalability.

Other researchers like Gordon and Catalini (2018) have highlighted the role of blockchain in enabling patient-centric control of medical data over institution-centric control. They have identified five mechanisms consisting of digital access rules, data aggregation, data liquidity, patient identity, and data immutability through which blockchain can address various challenges. Daisuke et al. (Ichikawa et al. 2017) proposed a Hyperledger fabric blockchain platform with a unique way to collect data using smartphones and introduced a concept called the "mHealth system." Stephen et al. (2018) identified the Ethereum platform as a potential candidate for managing healthcare data. Khezr et al. (2019) published a comprehensive review of the healthcare ecosystem, discussed some possible challenges of blockchain technology, and proposed an innovative delivery system called the internet of medical things (IoMT).

Currently, IBM have introduced "MiPasa" a global-scale control and communication system based on blockchain which manages COVID-19 outbreaks securely and democratically (Singh and Levi 2020). MiPasa utilizes blockchain technologies to gather reliable, quality data, address their inconsistencies, and make it easily accessible to technologists, data scientists, and public health officials. Omar et al. (2017) outlined a patient-centric healthcare data management system by using blockchain, which ensures privacy, integrity, accountability, and security of patients' data with cryptographic functions. Yue et al. (2016) proposed an app architecture called "Healthcare Data Gateway" (HGD) for personalized healthcare, enabling the patient to access, monitor, and manage their own data easily and securely store on a private blockchain. Zhou et al. (2018) proposed a blockchain-based medical insurance storage system and developed it on the Ethereum platform for patients, emergency clinics, insurance agencies, and servers. The proposed blockchain-based framework has three layers, namely user layer, system management layer, and storage layer to secure privacy and autonomy of the data. Studies related to managing patient records using the blockchain platform have made tremendous progress in recent times (Agbo et al. 2019; Khatoon 2020; Khezr et al. 2019).

Estonia is an outstanding example of the application of blockchain technology in the healthcare system (Aaviksoo 2020). During an emergency (such as a pandemic), a doctor can access critical information of patients, such as blood types, allergies, recent treatments, ongoing medication, or pregnancy. They deployed KSI® blockchain technology for the system to ensure data integrity and to mitigate internal threats. Ministries/government agencies can also utilize these systems to measure health trends, track epidemics, and ensure a better distribution of healthcare resources. A few key takeaways from this initiative are as follows: (i) blockchain can store patients' and clinical trial data for the development of an AI model to monitor, predict, diagnose, and treat COVID-19; (ii) blockchain can decentralize and democratize AI and machine learning model of COVID-19 to the rest of the world (Harris and Waggoner 2019); (iii) the blockchain can act like a central nervous system for an AI-linked network of sensors to predict important aspects of COVID-19 spread and outbreaks (Krittanawong et al. 2020); and (iv) with smart contract, blockchain can provide decentralized and autonomous pandemic organizations systems and AI can learn and adapt to changes over time to handle situations like COVID-19 pandemic without any delay and help regulatory agencies to take swift measures. Vulnerabilities of smart contracts can be detected by advanced techniques, further enhancing its security features (Tann et al. 2018; Gogineni et al. 2020).

5.4 ROLE OF AI IN DIAGNOSIS, DETECTION, AND IDENTIFICATION OF COVID-19 PATIENTS

In late 2002, there were three types of diagnostic tests standardized for SARS-CoV-1, namely: tissue culture isolation, antibody detection, and reverse transcription-polymerase chain reaction (RT-PCR) assays (Emery et al. 2004). However, the most widely used testing method for SARS-CoV-2 is RT-PCR (this is also by far the most reliable test). But this test has its own limitations in terms of time required for the analysis, availability in limited parts of the world, specimen collection, occasional poor performance, or in some cases, results are not reliable (Emery et al. 2004; Cheng 2020).

Meanwhile, it has been found that COVID-19 has particular signatures with lung CT scans, respiratory patterns, cough intensity, and others. Here, AI might play a crucial role in examining and detecting a particular signature and providing clinical decision support to doctors, specialists, and supporting staffs in real time. Lin Li (2020) developed a deep learning model to detect COVID-19 and to differentiate it from community-acquired pneumonia (CAP) and other lung diseases. The model extracts visual features from volumetric chest CT exams to detect COVID-19. A dataset from 4,356 chest CT exams was used and the model shows 90% and 96% per-exam sensitivity and specificity for detecting COVID-19 in the independent test, respectively. Also, Shuai Wang et al. (2020) used a deep learning algorithm to screen for COVID-19 using CT images. They deployed 1,119 CT images of datasets and modified the inception transfer-learning model to establish the algorithm. The model achieved a total accuracy of 89.5% with specificity of 0.88 and sensitivity of 0.87 in internal validation. Recently, Ophir Gozes et al. (2020) developed an AI-based automated CT image analysis tool for detection, quantification, and tracking of COVID-19. The study involves a test set consisting of 157 international patients (China and the United States) and classifies COVID-19 patients from others. They have reported two possible working points based on sensitivity and specificity: (i) 98.2% sensitivity and 92.2% specificity at high sensitivity point and (ii) 96.4% sensitivity and 98% specificity at high specificity point. Muhammad E. H et al. (Chowdhury et al. 2020) developed convolutional neural networks (CNNs) in screening viral and COVID-19 pneumonia. The database contains a mixture of digital X-ray images of 190 (COVID-19), 1,345 (viral pneumonia), and 1,341 (normal) cases. These networks were tested for two different classification schemes. The first scheme classifies normal and COVID-19 pneumonia patients with 98.3%, 96.7%, 100%, and 100% accuracy, sensitivity, specificity, and precision, respectively. The second scheme classifies normal, viral, and COVID-19 pneumonia patients with 98.3%, 96.7%, 99%, and 100% accuracy, sensitivity, specificity, and precision, respectively.

According to recent medical research, the patients infected with COVID-19 have more rapid respiration and have a different respiratory pattern than patients with the flu or common cold. It has been reported that the Tachypnea type of respiratory patterns has been observed in COVID-19 patients (Williamson 2020). Based on breathing characteristics, Yunlu Wang et al. (2020) deployed a deep learning model for prognosis, diagnosis, and screening of patients infected with COVID-19. The author utilized Gated Recurrent Unit (GRU) neural network with bidirectional and attentional method to classify six types of respiratory patterns such as Eupnea, Tachypnea, Bradypnea, Biots, Cheyne-Stokes, and Central-Apnea. M. Iqbal et al. (2020) proposed a robust method of active surveillance for COVID-19 patient using AI-based speech-recognition techniques through a mobile application to analyze cough sounds of suspected people and classify them in three levels as: mild, moderate, and severe. Li Yan et al. (2020) developed XGBoost machine learning-based COVID-19 prognostic prediction model using three clinical features, i.e., lactic dehydrogenase (LDH), lymphocyte, and high-sensitivity C-reactive protein (hs-CRP). The model used a dataset of 2,799 patients with more than 90% accuracy, enabling early detection of COVID-19. The remote monitoring and diagnosis of COVID-19 patients are essential to reduce the risk of transmission and also reduce hospital resource requirements during a pandemic (HIMSS Research 2020).

The amalgamation of travel history, recreational, and clinical and social data which includes the disease history of the family and lifestyle habits of individuals may be sourced from hospitals, travel websites, electronic media, etc., to enable more veracious predictions of an infected individual (Schootman et al. 2016). Skeptics may question that this may lead to a violation of the privacy of the masses. However, other AI applications have faced sufficient policy regulation that will corroborate the responsible use of this technology. The immediate diagnosis and preventive measures such as self-quarantine can be deployed so as to suppress the infection at a very nascent stage when it is easiest to control.

5.5 ACCELERATING DRUG DISCOVERY FOR COVID-19: THE BIG ROLE OF AI

The use of AI technologies is not limited to tracking, diagnostic, interpretation, and intervention of the coronavirus outbreak, but it is also playing an indispensable role in the discovery of COVID-19 drugs. Currently, structure-based drug designing approach is used to increase the speed of the drug designing process (Nero et al. 2018). The structure of proteins, i.e., the three-dimensional arrangement of atoms in an amino acid-chain molecule (Scheraga et al. 2006), is very important for designing a drug for a specific application (Kishan 2007). Nevertheless, determination of the protein structures experimentally is a difficult and intensive undertaking. Therefore, AI-based modeling approaches are also used for the prediction of protein structure for drug design. Recently, DeepMind has deployed its AlphaFold algorithm for understanding and predicting protein structures involved in SARS-CoV-2 induced infection and search for a potential source of treatment (Senior et al. 2020; Gulamali 2020). Senior et al. (2020) trained a neural network to obtain more information about the structure by accurate predictions of the distances between pairs of residues. This information is used to describe the shape of a protein by constructing a potential of mean force. Simple gradient descent algorithm is deployed to generate structures by optimizing the obtained potential.

Deep neural networks are also deployed to search for antivirals against new targets. The search is not limited to the experimental drugs alone but also applied to the library of approved drugs. Gene expression profiles of the molecular perturbations which are closely related to SARS-CoV-2 such as coatomer protein complex (COPB2) mainly composed the search domain (Avchaciov et al. 2020). The biomedical knowledge graphs can be used to discover drugs for SARS-CoV-2 infection. Some of the well-known examples of knowledge graphs include Freebase (Bollacker et al. 2008), DBpedia (Bizer et al. 2009), Nell (Carlson et al. 2010), and YAGO (Hoffart et al. 2013). The biomedical knowledge graph is made up of a vast number of structured medical information such as diseases, genes, and drugs and with over

20 types of biomedical entities and their relationship extracted from the scientific literature by machine learning (Kamilaris et al. 2019; Sang et al. 2018). Benevolent AI's knowledge graph is utilized to search for approved drugs that could be of help for COVID-19 (Richardson et al. 2020) as depicted in Figure 5.2. The author identified baricitinib as a potential treatment for COVID-19, which can be utilized to reduce the ability of the virus to infect lung cells.

Further, AI models can be utilized to predict COVID-19 protein-ligand binding affinities in order to identify some potential old drugs. A multitask deep model is used to predict potential commercial inhibitors against SARS-CoV-2 by screening the available drug dataset of 4,895 samples. The binding affinity (pKa) between drug and target is the output of the model. The results show that the abacavir inhibitor, used for the treatment of HIV, has a high binding affinity with several proteins of SARS-CoV-2. The model also predicted ten drugs as potential inhibitors.

Also, biotech and AI firms are increasingly coming together to use and provide AI-based platforms to assist the drug discovery community, for example, InSilico Medicine and DeepMind joining together and providing AI/ML pipelines for drug discovery research (Avchaciov et al. 2020). Pharmaceutical industries such as Pfizer, Amgen, and Sanofi are also using a blockchain-based digital drug control system (DDCS) to store information about the inspection and evaluation of new drugs (Markov 2018).

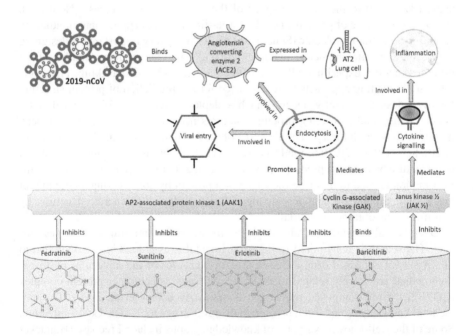

FIGURE 5.2 The use case of AI-based biomedical knowledge graph to search and identify the approved drugs that could help for COVID-19; the results show baricitinib as the potential treatment for 2019-nCoV (Richardson et al. 2020).

Recently, Exscientia has tied up with Diamond Light Source (the UK's national synchrotron facility) and Calibr (a division of Scripps Research, USA) to discover potential viable drugs for COVID-19 using their AI-based expertise (Crisp and Camille 2020). Exscientia will screen the key viral drug targets of SARS-CoV-2 virus using its advanced biosensor platforms. Some of these critical targets include the virus's spike protein, the 3CL protease, and the NSP 12-NSP7-NSP8 RNA polymerase complex. After finding a viable drug through the screening process, the start-up will move toward designing superior AI-powered drug molecules for COVID-19. Earlier this year, they designed a drug molecule using AI for the treatment of obsessive compulsion disorder, thus becoming the first AI-based drug to enter human clinical trials in the history of medicine (Murgia 2020). Considering all these successes in AI-based drug discovery, it is reasonable to be hopeful of achieving a positive result sooner than later.

5.6 SECURING DIFFERENT SUPPLY CHAINS AT TIME OF CRISIS

5.6.1 Medical Supply Chain

The COVID-19 crisis has put tremendous pressure on the existing infrastructure of medical supply chains around the world. Critical medical supplies such as ventilators, testing kits, and personal protection equipment (PPE) are running in short supply worldwide. To address this, contracts are being signed to upscale manufacturing worth billions of dollars. Key obstacles faced by different stakeholders are trust-deficit challenges such as the credibility of suppliers, diverse product requirements and standards, upfront financial payments, custom certifications, and tracking of transporting vehicles. These issues can be easily addressed through the implementation of blockchain across the healthcare system. Various studies using blockchain platforms have been conducted to track fraudulent drug supplies. Sylim et al. (2018) proposed the design for a pharmacosurveillance blockchain system to improve the monitoring of fraudulent medical supplies. This surveillance system will be based on Ethereum as well as a Hyperledger fabric platform. Tseng et al. (2018) developed a Gcoin based blockchain, where the letter G in Gcoin represents Global Governance in blockchain network. This scheme was designed to create a transparent drug supply chain among various stakeholders. Recently, IBM launched "Trust Your Supplier" (Khurshid 2020), a blockchain-based platform to ensure transparency among different nodes on the supply chain such as buyer, manufacture, and seller as shown in Figure 5.3. Although we are at the early stages of developing blockchain solutions for our medical supply chain system, based on recent progress in literature, there are reasons to be hopeful about its implementation shortly.

5.6.2 Food Supply Chain

Due to the worldwide lockdown during the pandemic situation, food industries are struggling, local suppliers and farmers are terrified, and with economic activity already declining, this might lead to a food crisis. The heads of three global

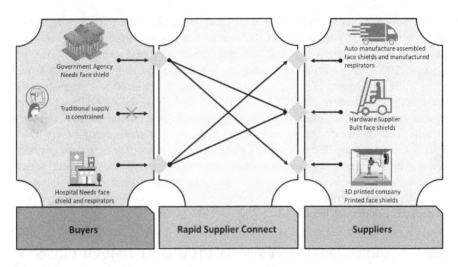

FIGURE 5.3 Schematic of IBM "Rapid Supplier Connect" for buyers, distributors, sellers, or suppliers.

agencies, namely the Food and Agriculture Organization (FAO), World Health Organization (WHO), and World Trade Organization (WTO), also warned of the risk of a worldwide "food shortage" if the COVID-19 pandemic is not properly managed by the concerned authorities (Malsang 2020). Tracing and monitoring illegal storages, foodborne illness, food recall/loss/fraud, and food availability/access and identifying other weak links are very difficult during a pandemic (Koonce 2017). However, blockchain may provide a credible solution to these problems by streamlining the food supply chains (Kamilaris et al. 2019). Fran Casino et al. (2019) have developed a fully functional smart contract and a local private blockchain-based distributed, decentralized, and automated functional model to ensure food supply chain traceability. M. Caro et al. (2018) designed "AgriBlockIoT," a decentralized supply chain management system based on Internet of Things (IoT) and blockchain technology to improve the traceability in agri-food supply chains. Dianhui Mao et al. (2018) used consortium blockchain technology to designs a novel Food Trading System with COnsortium blockchaiN (FTSCON) to put an end to information asymmetry in the food supply chain using online double auction mechanism and an improved Practical Byzantine Fault Tolerance (iPBFT) algorithm. Walmart partnered with IBM to create a blockchain-based decentralized food traceability system based on Hyperledger fabric to trace mangoes sold in Walmart's US stores and pork sold in its China stores (Hyperledger 2020). In the United States, Cybersecurity and Infrastructure Security Agency (CISA) classified distributed ledger blockchain technology for agriculture and food supply chain in COVD-19 (Thind 2020). IBM is using a product called "IBM Food Trust" based on blockchain to improve transparency in food supply chains during the pandemic (Reilly 2020).

5.6.3 RELOCATION OF MEDICAL, LAW ENFORCEMENT, AND OTHER CRITICAL SERVICE PROFESSIONALS

In general, pandemics spread from region to region just like wildfires. SARS-CoV-2 is no exception (Costantino et al. 2020); being a contagious disease, it spreads via contact, coughing, sneezing, etc. (J. Wang and Du 2020). The time series model can be utilized to forecast the rate of increase of cases for the next few weeks (Weigend 1994). Ensemble learning algorithms can be utilized to enhance the accuracy of the prediction (Giovanni and Elder 2010). Furthermore, blockchain has proved to be effective in crisis management situations such as disaster management (Demir et al. 2018). So, extrapolating its effects to contain a pandemic is a logical extension of the idea and should be seriously evaluated. It can provide vital data to the governments which, if properly used for data mining, can provide vital clues and recommendations as to "how to contain the virus." It can act as a platform where all the concerned authorities, such as medical professionals, governments, media, researchers, and others, can update each other about the situation and avert an impending danger.

5.7 IMPACT OF COVID-19 ON WORLD ECONOMY AND ROLE OF AI AND BLOCKCHAIN

With the outbreak of COVID-19 and the consequent lockdown of billions of people around the world, the economies of many countries are going into recession. Due to the lack of demand and various supply chain constraints, this outbreak has spiraled into cascading economic disruptions across different continents. S&P, a major global rating agency, predicted a sharp contraction of 3.8% in the global growth for the year 2020 (Gruenwald 2020). In the United States, the unemployment rate as of July 2020 was at a multi-decade high of 10.2%, surpassing the previous record observations during the global financial crisis (GFC) with a shocking Q2 GDP contraction of 32.9% (Dock David Treece 2020; Duffin 2020). The GDP growth in India contracted to a multi-decade low of 23.9% in the first quarter of fiscal year 20–21 (Reuters 2020). India's unemployment rate also touched more than 20% during the peak of the pandemic in April 2020 (CMIE Database 2020). China had reported severe contractions across different economic indicators such as Purchasing Managers' Index (PMI) industrial output, fixed investment, and retail sales. VIX volatility index, a gauge for fear in US stock markets, reached its highest value since the global financial crisis during the crash and ultimately destabilized the global financial markets for a certain period of time. However, massive monetary interventions by central banks around the world and a historical low in ten-year US bond yields have contributed to a spectacular V shape recovery across major global stock indices, with S&P 500 so far witnessing a 53% rise from March lows.

To tackle this impact of COVID-19 on the economy, central banks and governments around the world are acting in a coordinated manner to spur demand and to make cash available in the hands of people and businesses by announcing economic stimulus packages to the tune of trillions of dollars. For example, the United States

Federal government has signed a 2.2 trillion USD financial package for vulnerable sections of society and businesses hit hardest by this pandemic (Mascaro 2020). The union government of India also announced an economic stimulus of over 22 billion USD, also specifically targeting economically weaker sections of the country (Choudhury 2020).

The following sections will throw some light on how blockchain technology might help the world to come out of this economic recession in a lesser time under these prevailing conditions. To put the argument in context, first we depict the possible causes of the global financial crisis in 2008 using a broad stroke and understand how blockchain technology could have helped to prevent such an event. The financial crisis in 2008 started due to risky subprime mortgaged home loans distributed to people who could hardly afford them. These unsafe lending practices were prevalent not only in the United States but among major economies. These practices resulted in a series of cascading events starting from default in loan payments, seizure of houses by authorities, and finally selling of properties by lenders to recover their dues. This led to the flooding of houses in real estate markets, thus damaging the prices of houses drastically. It caused panic among home loan borrowers, and many of them defaulted willingly to take advantage of lower home prices. Finally, it led to large chunks of defaults by home loan borrowers and bankruptcy of a prominent investment bank, Lehmann Brothers, thus creating a void of trust among financial institutions. The aftermath of the fallout of such an event created chaos among financial institutions, resulting in a severe credit crunch in various economies around the world. The main culprit for the lack of liquidity in financial markets was the issue of trust among the bankers at that crucial point. These institutions failed to collectively take rational direction under those uncertain circumstances and pulled the trigger to call out all the outstanding dues. But this situation could have been better handled through the implementation of blockchain technology and, similarly by extension, great catastrophes in future can be avoided by putting in place a robust mechanism to ensure trust even in difficult times and create a conducive environment where collective rational decisions can be made. Very crudely, the primary reason for such a crisis was the lack of liquidity in the system. Using blockchain protocols throughout the entire industry, banks might avoid a credit crunch situation in the future. The present means of transactions in banks is through creating agreements between two parties, which not only carries a default risk on one party, but is also time-consuming. Both of these problems can be largely resolved through the integration of smart, automated contracts on the blockchain platform, thus taking care of trust issues among bankers.

The present economic crisis arising out of COVID-19 is entirely different from the global financial crisis in 2008. The current turmoil is more of a supply chain issue due to the lockdown of almost the entire population. However, if this lockdown continues for more time, it will have a severe impact on businesses. Industries and people will soon run out of financial resources at their disposal, thus creating a situation not very different from the credit crunch in markets. Central banks around the world are reducing interest rates and taking appropriate measures to induce liquidity in financial markets. But these measures take time to be effective, thus prolonging this current crisis, and the responses are often not coordinated and therefore do not yield the best possible results.

On the other hand, AI can play a vital role and drive economic change by harnessing new tools such as computer vision, natural language, virtual assistants, robotic process automation, and advanced machine learning to drive the new economic growth. Accenture (Accenture Research 2020) reported that AI could double the annual economic growth of 12 developed economies by 2035. Similarly, PricewaterhouseCoopers (PwC) reported that AI could increase the global GDP to 14% by 2030 (Berger 2018).

However, every crisis always gives rise to new technologies, innovations, leaders, and above all, opportunities. Perhaps, this crisis will allow policymakers to rethink their strategy about blockchain technology and implement/integrate it into the existing financial system to make it even more robust.

5.8 OTHER SMART INTERVENTIONS

5.8.1 FACIAL RECOGNITION AND FEVER DETECTOR AI

Currently, positive cases for COVID-19 are increasing globally, and on the other hand, there is lack of testing kits and personal protective equipment for clinical staffs, further complicating the overall situation. Other detection systems such as infrared ear thermometers work for one person at a time in a close range which may allow the transmission of infection to another person. Therefore, facial recognition and thermal imaging integrated with AI can play a vital role in this regard by scanning several hundred people without any contact. AI-integrated thermal imaging techniques are already well known in healthcare applications, particularly for the diagnosis of breast cancer due to their fast, non-invasive, non-contact, and flexible nature (Mambou et al. 2018). Austin-based Athena Security developed thermal imaging and AI-based cameras to detect the people infected with COVID-19 in public and other crowded places like airports, grocery stores, hospitals, and medical regions. The cameras can take 1,000 temperature readings in an hour, with an accuracy of within half a degree (Topol 2020). Miso Robotics, a start-up, is deploying robots and AI screening devices at doors to detect, scan, and take the body temperature of staff, delivery drivers, and guests with COVID-19 virus (Wiggers 2020). The security at Hangzhou park, in eastern China, is using AI-powered smart glasses with non-contact thermal augmented reality to detect and measure the body temperature of park visitors who have fever possibly related to the virus, keeping a distance of up to 1 meter (Dai 2020).

5.8.2 AI-ENABLED DRONES AND ROBOTS

Intelligent drones, swarms of coordinated drones, and robots hold the ability and great promise to flatten the infection curve. Drones can be connected to a ground-based server or with in-built natural language processing systems, speech recognition, predictive analysis, and others can be used for contact tracing. Intelligent drones can also be used for disease monitoring (Rosser et al. 2018). The areas which are most affected can be identified; the number of medical staff and medical equipment required can be noted and deployed swiftly. These drones can be even more effective in tracing asymptomatic patients and overseeing their activities (Jain et al. 2018). Many companies in the world are coming forward to develop pandemic drones, which will remotely monitor and discern individuals or groups of individuals with infectious conditions or respiratory ailments. Sensors, cameras, and other computer vision systems can be fitted which would keep track and regulate body temperature, heart rate, respiratory rates, etc. They can also be programmed to detect those who are sneezing and coughing among a congregation (Cozzens 2020).

The COVID-19 pandemic spreads through contact and the virus is said to thrive on different surfaces for 24–72 hours (Davis 2020). Humans touching these surfaces risk infecting themselves with the virus. So, robots can come in handy in taking samples from infected surfaces and analyzing them further (Pajares 2015). Patients' care, averting the risk to healthcare workers, has also benefited from the increasing use of robots for the delivery of food and medicine. The job of cleaning and sanitation of isolation wards is now being entrusted to robots (Yu et al. 2018). All these, in turn, have the power to decrease the disease burden of countries/communities, thereby making the world safer.

As the war against epidemics becomes preeminent, versatile technologies are needed to perceive and diagnose outbreaks. This would lead to more expeditious implementations of critical interventions with more coherence.

5.8.3 Virtual Healthcare Assistants (Chatbots)

The rampant increase of COVID-19 cases across the world has shown that contemporary healthcare systems and mitigation measures can be overwhelmed. Natural language processing capabilities can play an important role in managing the burden on healthcare systems during such pandemics by building a multilingual virtual healthcare agent that will be able to answer questions related to COVID-19, propose protection measures, provide dependable information (Amato et al. 2017) and comprehensible guidelines, examine symptoms, undergo constant monitoring, and notify individuals about the requirements of hospital-based quarantine or home quarantine. These chatbots may be fed with tailored algorithms based on different factors such as travel history, current climatic conditions, and past disease records (Denecke et al. 2019; W. Wang and Siau 2018) and can answer the queries of the individuals while they are at their homes via the internet. This is most advantageous in handling contagious viruses as it reduces the need for visiting medical facilities, thereby reducing the chance of exposure and infection. However, expert advice from healthcare professionals must be taken while developing/deploying/operating these chatbots. The diagnosis provided by these chatbots should be thoroughly monitored. Different companies are coming up with interactive voice and text chatbots, which are capable of educating the public and assist in triage (Schwartz 2020).

Virtual healthcare assistants can succor not only individuals but also companies. They can delineate different methods to make working from home effective by analyzing the types of queries. On one hand, it can facilitate employers to monitor the progress of the work, and on the other hand, it can facilitate employees to reach out to their team quickly and resolve their queries (Kar and Haldar 2016).

5.8.4 Managing Finance/Donation (Blockchain)

Blockchain is a record of events that can be shared and accessed by a network of computers. It is managed by a peer-to-peer network that adheres to a protocol for inter-node communication and inclusion of new blocks. They are considered secure by design because the data in any given block cannot be altered retroactively without the alteration of all subsequent blocks (Zīle and Strazdiņa 2018).

Various non-governmental organizations, individuals, businesses, etc., are coming forward to donate money to ensure that proper medical care reaches the ailing patients. However, it is not always obvious, if at all, whether the money that has come as an aid for the patients has reached the beneficiary or not. Blockchain can be effectively used to resolve this dilemma. Additionally, if a donor can see that their donation is reaching a needy person, then that will create tremendous psychological satisfaction and further encourage more donations in the future. This will also enable donors to perceive where the requirement of funds is vital, and they can then channel their money into such places (Saleh et al. 2019). Also, they can track their donation till it reaches its destination. Blockchain thus provides the general public with lucid information as to how their donations are being utilized and will further encourage more philanthropy.

5.9 OUTLOOK/PERSPECTIVE

5.9.1 Trust: The Missing Piece in Today's Societal and Technological Architecture

Philosophically speaking, the very fabric of human civilization is established on trust. It is an essential protocol of transaction between two organizations, countries, or even in a marriage or relationship. However, this very social fabric becomes severely strained during a big crisis. A global pandemic, such as the present ongoing one caused by SARS-CoV-2, is the best example for it. In the absence of a vaccine or medicine to cure it, the best humankind can do is mobilize resources to the fullest extent to fight against this common and brutal enemy. But it was observed that in some cases, different countries and even different provinces/states were fighting with each other, while instead they should have focused on fighting this invisible enemy that has created an existential threat for the entirety of humankind. To a great extent, this is a reflection of trust deficit leading to competition and conflict rather than cooperation. This chapter has therefore outlined how blockchain can be harnessed not only during the time of crisis but as a way of building a robust system that will not yield even during such a crisis and will keep us prepared to face similar and bigger challenges.

5.9.2 INTEGRATED AI AND BLOCKCHAIN ECOSYSTEM: SMART DEFENSE AGAINST GLOBAL PANDEMICS

Over the last decade, the face of human civilization has been completely altered by AI. The availability of very cheap computing resources, primarily because of the huge development of graphics processing units (GPUs) and major breakthroughs in machine learning, particularly the advent of deep learning, has made it possible. In fact, AI holds tremendous potential to dramatically accelerate critical researches like drug discovery and development of vaccines. However, the vast majority of these efforts are fragmented. Most likely, during the post-COVID-19 situation, when we will observe this retrospectively, we will realize how our responses were severely constrained because of this fragmentation. We will therefore make a serious attempt to properly integrate our efforts. Also, it has been discussed that AI thrives on a large amount of authentic data, which is a forte of blockchain technology. Therefore, we envision that COVID-19 will catalyze a much closer integration of these two technologies. Another piece of critical technology that we feel will gain huge prominence and have huge importance in our future endeavors is quantum computing. We did not discuss it at length in this chapter because it is not yet mature enough to make a huge impact on the ongoing pandemic. However, we have no doubt that in the future it will play a big role in drug discovery, vaccine development, managing logistics, etc. We also envision huge interest in quantum computing in the post-COVID-19 world.

One of the prime reasons for the failure of our technical undertaking that stands out too much is the centralized nature of our technological architecture, which prohibited smooth communication between different animate or inanimate agents that are part and parcel of this war. Blockchain, as a decentralized tool, can address that concern, as well as helping to mitigate a large chunk of problems faced during this emergent situation. The implementation of blockchain technology in this situation makes it feasible to utilize the data present across a multitude of industries and to process it (Nichol et al. 2016). Public health data, particularly for infectious disease outbreaks such as COVID-19, can be tracked across countries using blockchain. As transparency increases, there can be a significant increase in the accuracy of reporting and receiving efficient responses. Blockchain allows for expeditious processing of authentic data, thereby facilitating identification of patients through early detection of symptoms (Talukder et al. 2018).

5.10 FUTURE RESEARCH DIRECTIONS

5.10.1 THE BIG CANVAS OF INTEGRATION OF EMERGING TECHNOLOGICAL PARADIGMS

There is a wide range of digital technologies beyond AI and blockchain, such as the Internet of Things (IoT), augmented reality and virtual reality (AR/VR), big data, and quantum computing, as shown in Figure 5.4. These digital technologies can be merged with AI and blockchain to form a strong and smart first line of defense against any future pandemic. This emerging system can do myriads of chores swiftly,

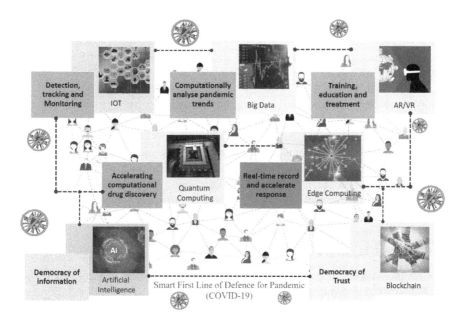

FIGURE 5.4 Smart first line of defense against pandemic with different digital technologies such as the Internet of Things (IoT), augmented reality and virtual reality (AR/VR), big data, and quantum computing.

such as managing innovation and drug discovery, tracking donations, ensuring the supply of essential commodities and medical supplies, managing medical data, overseeing the drug supply chains, warning citizens, maintaining the traveling history of individuals, even managing public health strategies, and many more, a majority of which are sketched in this chapter. Many of these ideas might even be beneficial in containing the disease as it unfolds in front of our eyes, before it spreads to the level of a great crisis.

5.10.2 Development of Integrated Frameworks

The present chapter is enough to convince us that a huge amount of research works is being undertaken that can form the basis of our fight against future global pandemics. However, what is seriously lacking is as follows: (i) An integrated technological framework that can put the diverse computational infrastructure in a systematically integrated framework. Two analogies may be of interest here. The first analogy involves Hadoop and, in particular, the way it did try to create a software framework for distributed storage and processing of "Big data" using MapReduce, which was originally discovered by Google. There is an urgent need to develop a similar integrated framework for the model space, though it might be much tougher. The second analogy is about Integrated Computational Materials Engineering (ICME) framework, which aims to integrate different computational materials science models that are poles apart in their capabilities in handing the system size, time, and length scales. (ii) A collective will of policymakers and politicians to drive such a metamorphic technological transformation. Humanity has witnessed how nearly unthinkable scientific projects can be successfully pursued if there is a clear intent. The Human Genome Project, CERN, and ITER are some ready examples. The scientific community therefore should seriously debate how to create a truly global initiative in preparation for future pandemics. (iii) Participation of common people can do wonders. This is one critical area that should never be overlooked. No matter how sophisticated our technological architecture is, it will never be successful without decisive and active participation by all. Open research, crowdfunding, and crowdsourcing have started but their reach is still insignificant and a lot more new initiatives are necessary. The entire community should seriously introspect how to build the bridge from both ends.

5.11 CONCLUSION

COVID-19 disease has astounded the world and has shaken the core of human psychological, economic, and technical frameworks. The wings of trepidation have gripped the entirety of humankind. A very sincere attempt has been made in this chapter for a serious introspection of our existing technological architecture. In particular, we analyzed why it was so awfully inadequate despite its astounding success in our daily life. However, not all is gloomy, and the crisis has brought out some good aspects to the forefront. The healthcare sector, business communities, and governments across the world have largely joined together in stifling the spread of the disease. By creating "trust" in the digital space, we can produce a nearly flawless, highly integrated technology architecture, where two dissimilar inanimate agents can engage with each other. This might form the fundamental building blocks for our robust first line of defense to fend off a similar or bigger crisis in the future. One of the most prominent candidate technologies for this is blockchain. Its effectiveness can be very significantly enhanced by coupling it with AI. AI and blockchain are coming out in a big way, and they will likely take a lead role in the technological arena in the near future. We reasoned how the combination of AI and blockchain technology could help us overcome these critical shortcomings. Since AI thrives on a large amount of trustworthy data, which will be a natural outcome of using blockchain as a "connective tissue," together they will make a very potent combination that will stand tall even in the face of a big crisis like COVID-19 or even a bigger one. Thus, they will feed one another and will create a very strong symbiotic relationship that will create our robust first line of defense against any such future challenges. In particular, this chapter discusses some emergent directions that can be effectively addressed by these two technologies combined (AI and blockchain): accelerating drug discovery, managing medical and food supply chains, resource allocations in critical times, transparent management of donations, limiting the harm on economics, smart use of coordinated swarms of drones and robots for effective deployment during the crisis, and last but not least, possible routes for curing COVID-19.

REFERENCES

Aaviksoo, Ain. 2020. "Building Blockchain Powered Trusted Digital Health Services. Estonia." *Blockchain in Healthcare Today* 3 (January). https://blockchainhealthcaretoday.com/index.php/journal/article/view/129.

Abhari, Reza S., Marcello Marini, and Ndaona Chokani. 2020. "COVID-19 Epidemic in Switzerland: Growth Prediction and Containment Strategy Using Artificial Intelligence and Big Data." *MedRxiv*, https://doi.org/10.1101/2020.03.30.20047472.

Accenture Research 2020. "Artificial Intelligence Is the Future of Growth." https://www.accenture.com/in-en/insight-artificial-intelligence-future-growth. (accessed April 4, 2020).

Agbo, Cornelius, Qusay Mahmoud, and J. Eklund. 2019. "Blockchain Technology in Healthcare: A Systematic Review." *Healthcare* 7 (2): 56. https://doi.org/10.3390/healthcare7020056.

Amato, Flora, Stefano Marrone, Vincenzo Moscato, Gabriele Piantadosi, Antonio Picariello, and Carlo Sansone. 2017. "Chatbots Meet Ehealth: Automatizing Healthcare." *CEUR Workshop Proceedings* 1982: 40–49.

Avchaciov, Konstantin, Olga Burmistrova, and Peter Fedichev. 2020. *AI for the Repurposing of Approved or Investigational Drugs against COVID-19.* https://doi.org/10.13140/RG.2.2.20588.10886.

Berger, Irving Wladawsky. 2018. "The Economic Value of Artificial Intelligence." https://blog.irvingwb.com/blog/2018/10/the-economic-value-of-artificial-intelligence-.html. (accessed April 4, 2020).

Bizer, Christian, Jens Lehmann, Georgi Kobilarov, Sören Auer, Christian Becker, Richard Cyganiak, and Sebastian Hellmann. 2009. "DBpedia - A Crystallization Point for the Web of Data." *Journal of Web Semantics* 7 (3): 154–65. https://doi.org/10.1016/j.websem.2009.07.002.

Bollacker, Kurt, Colin Evans, Praveen Paritosh, Tim Sturge, and Jamie Taylor. 2008. "Freebase: A Collaboratively Created Graph Database for Structuring Human Knowledge." In *Proceedings of the 2008 ACM SIGMOD International Conference on Management of Data*, 1247–50. SIGMOD '08. New York: Association for Computing Machinery. https://doi.org/10.1145/1376616.1376746.

Carlson, Andrew, Justin Betteridge, Bryan Kisiel, Burr Settles, Estevam R. Hruschka, and Tom M. Mitchell. 2010. "Toward an Architecture for Never-Ending Language Learning." *Proceedings of the National Conference on Artificial Intelligence* 3: 1306–13.

Caro, M P, M S Ali, M Vecchio, and R Giaffreda. 2018. "Blockchain-Based Traceability in Agri-Food Supply Chain Management: A Practical Implementation." In *2018 IoT Vertical and Topical Summit on Agriculture - Tuscany (IOT Tuscany)*, 1–4. https://doi.org/10.1109/IOT-TUSCANY.2018.8373021.

Casino, Fran, Venetis Kanakaris, Thomas K Dasaklis, Socrates Moschuris, and Nikolaos P Rachaniotis. 2019. "Modeling Food Supply Chain Traceability Based on Blockchain Technology." *IFAC-PapersOnLine* 52 (13): 2728–33. https://doi.org/10.1016/j.ifacol.2019.11.620.

Cennimo, David J. 2020. "Coronavirus Disease 2019 (COVID-19) Treatment & Management." https://emedicine.medscape.com/article/2500114-treatment (accessed April 6, 2020).

Cheng, Xu. 2020. "Huiying Medical: Helping Combat COVID-19 with AI Technology." https://www.intel.com/content/www/us/en/artificial-intelligence/posts/huiying-medical-covid19.html (accessed April 8, 2020).

Choudhury, Saheli Roy. 2020. "India Announces $22.5 Billion Stimulus Package to Help Those Affected by the Lockdown." https://www.cnbc.com/2020/03/26/coronavirus-india-needs-a-support-package-larger-than-20-billion-dollars.html (accessed April 4, 2020).

Chowdhury, Muhammad E.H., Tawsifur Rahman, Amith Khandakar, Rashid Mazhar, Muhammad Abdul Kadir, Zaid Bin Mahbub, Khandakar R. Islam, et al. 2020. "Can AI Help in Screening Viral and COVID-19 Pneumonia?," http://arxiv.org/abs/2003.13145.

CMIE Database 2020. "Unemployment Rate in India." https://unemploymentinindia.cmie.com/ (accessed October 18, 2020).

Costantino, Valentina, David Heslop, and Chandini MacIntyre. 2020. "The Effectiveness of Full and Partial Travel Bans against COVID-19 Spread in Australia for Travellers from China." *MedRxiv* 21 (1): 1–9. https://doi.org/10.1101/2020.03.09.20032045.

Cozzens, Tracy. 2020. "Pandemic Drones to Monitor, Detect Those with COVID-19." https://www.gpsworld.com/draganfly-camera-and-uav-expertise-to-help-diagnose-coronavirus (accessed April 4, 2020).

Crisp, Stephanie, and Oster Camille. 2020. "Exscientia Announces Joint Initiative to Identify COVID-19 Drugs with Diamond Light Source and Scripps Research." https://www.exscientia.ai/news-insights/exscientia-announces-joint-initiative-to-identify-covid-19 (accessed April 10, 2020).

Dai, Sarah. 2020. "Covid-19: Hangzhou Park Security Uses AI-Powered Smart Glasses to Detect People with Fever." https://www.thestar.com.my/tech/tech-news/2020/03/27/covid-19-hangzhou-park-security-uses-ai-powered-smart-glasses-to-detect-people-with-fever (accessed April 4, 2020).

Davis, Nicola. 2020. "How Long Does Coronavirus Live on Different Surfaces?" https://www.theguardian.com/world/2020/may/27/how-long-does-coronavirus-survive-on-different-surfaces (accessed April 4, 2020).

Demir, M, A.A. Mashatan, O. Turetken, and A. Ferworn. 2018. "Utility Blockchain for Transparent Disaster Recovery." In *2018 IEEE Electrical Power and Energy Conference (EPEC)*, 1–6. https://doi.org/10.1109/EPEC.2018.8598413.

Denecke, Kerstin, Mauro Tschanz, Tim Lucas Dorner, and Richard May. 2019. "Intelligent Conversational Agents in Healthcare: Hype or Hope?" *Studies in Health Technology and Informatics* 259 (April): 77–84. https://doi.org/10.3233/978-1-61499-961-4-77.

Dock David Treece. 2020. "How Bad Was Second Quarter GDP, Really?" *Forbes*. https://www.forbes.com/sites/advisor/2020/07/31/how-bad-was-second-quarter-gdp-really/#1254c5fd4430 (accessed April 1, 2020).

Duffin, Erin. 2020. "Monthly Unemployment Rate in the United States from July 2019 to July 2020." *Statista*. https://www.statista.com/statistics/273909/seasonally-adjusted-monthly-unemployment-rate-in-the-us/.

Emery, Shannon L, Dean D Erdman, Michael D Bowen, Bruce R Newton, Jonas M Winchell, Richard F Meyer, Suxiang Tong, et al. 2004. "Real-Time Reverse Transcription–Polymerase Chain Reaction Assay for SARS-Associated Coronavirus." *Emerging Infectious Disease Journal* 10 (2): 311. https://doi.org/10.3201/eid1002.030759.

Fong, Simon James, Gloria Li, Nilanjan Dey, Rubén Gonzalez-Crespo, and Enrique Herrera-Viedma. 2020. "Finding an Accurate Early Forecasting Model from Small Dataset: A Case of 2019-NCoV Novel Coronavirus Outbreak." *International Journal of Interactive Multimedia and Artificial Intelligence* 6 (1): 132. https://doi.org/10.9781/ijimai.2020.02.002.

Giovanni, Seni, and John F. Elder. 2010. *Ensemble Methods in Data Mining: Improving Accuracy Through Combining Predictions*. Edited by Robert Grossman. 1st ed. Morgan & CLaypool Publishers. https://doi.org/10.2200/S00240ED1V01Y200912DMK002.

Gogineni, Ajay K., S. Swayamjyoti, Devadatta Sahoo, Kisor K. Sahu, and Raj Kishore. 2020. "Multi-Class Classification of Vulnerabilities in Smart Contracts Using AWD-LSTM, with Pre-Trained Encoder Inspired from Natural Language Processing." http://arxiv.org/abs/2004.00362.

Gordon, William J., and Christian Catalini. 2018. "Blockchain Technology for Healthcare: Facilitating the Transition to Patient-Driven Interoperability." *Computational and Structural Biotechnology Journal* 16: 224–30. https://doi.org/10.1016/j.csbj.2018.06.003.

Gozes, Ophir, Maayan Frid-Adar, Hayit Greenspan, Patrick D. Browning, Adam Bernheim, and Eliot Siegel. 2020. "Rapid AI Development Cycle for the Coronavirus (COVID-19) Pandemic: Initial Results for Automated Detection & Patient Monitoring Using Deep Learning CT Image Analysis." http://arxiv.org/abs/2003.05037.

Gruenwald, Paul F. 2020. "Economic Research: The Global Economy Begins A Slow Mend As COVID-19 Eases Unevenly." *S&P Global Ratings*. https://www.spglobal.com/ratings/en/research/articles/200701-economic-research-the-global-economy-begins-a-slow-mend-as-covid-19-eases-unevenly-11552670 (accessed August 4, 2020).

Gulamali, Faris. 2020. "AlphaFold Algorithm Predicts COVID-19 Protein Structures." https://www.infoq.com/news/2020/03/deepmind-covid-19/ (accessed April 10, 2020).

Harris, Justin D., and Bo Waggoner. 2019. "Decentralized and Collaborative AI on Blockchain." In *Proceedings – 2019 2nd IEEE International Conference on Blockchain, Blockchain 2019*, 368–75. https://doi.org/10.1109/Blockchain.2019.00057.

HIMSS Research 2020. "Remote Patient Monitoring: COVID-19 Applications and Policy Challenges." https://www.himss.org/news/remote-patient-monitoring-covid-19 (accessed August 4, 2020).

Hoffart, Johannes, Fabian M. Suchanek, Klaus Berberich, and Gerhard Weikum. 2013. "YAGO2: A Spatially and Temporally Enhanced Knowledge Base from Wikipedia." *Artificial Intelligence* 194: 28–61. https://doi.org/10.1016/j.artint.2012.06.001.

Hu, Zixin, Qiyang Ge, Shudi Li, Li Jin, and Momiao Xiong. 2020. "Artificial Intelligence Forecasting of Covid-19 in China." 1–20. http://arxiv.org/abs/2002.07112.

HyperLedger. 2020. "How Walmart Brought Unprecedented Transparency to the Food Supply Chain with Hyperledger Fabric." *Case Study*. https://www.hyperledger.org/resources/publications/walmart-case-study (accessed August 4, 2020).

Ichikawa, Daisuke, Makiko Kashiyama, and Taro Ueno. 2017. "Tamper-Resistant Mobile Health Using Blockchain Technology." *JMIR mHealth and uHealth* 5 (7): 1–10. https://doi.org/10.2196/mhealth.7938.

Iqbal, Mohammad. 2020. *Active Surveillance for COVID-19 through Artificial Intelligence Using Concept of Real-Time Speech-Recognition Mobile Application to Analyse Cough Sound*. https://doi.org/10.31219/osf.io/cev6x.

Jain, Trevor, Aaron Sibley, Henrik Stryhn, and Ives Hubloue. 2018. "Comparison of Unmanned Aerial Vehicle Technology versus Standard Practice in Identification of Hazards at a Mass Casualty Incident Scenario by Primary Care Paramedic Students." *Disaster Medicine and Public Health Preparedness* 12 (5): 631–34. https://doi.org/10.1017/dmp.2017.129.

Kamilaris, Andreas, Agusti Fonts, and Francesc X. Prenafeta-Boldύ. 2019. "The Rise of Blockchain Technology in Agriculture and Food Supply Chains." *Trends in Food Science & Technology* 91: 640–52. https://doi.org/10.1016/j.tifs.2019.07.034.

Kar, Rohan, and Rishin Haldar. 2016. "Applying Chatbots to the Internet of Things: Opportunities and Architectural Elements." *International Journal of Advanced Computer Science and Applications* 7 (11). https://doi.org/10.14569/ijacsa.2016.071119.

Khatoon, Asma. 2020. "A Blockchain-Based Smart Contract System for Healthcare Management." *Electronics (Switzerland)*. https://doi.org/10.3390/electronics9010094.

Khezr, Seyednima, Md Moniruzzaman, Abdulsalam Yassine, and Rachid Benlamri. 2019. "Blockchain Technology in Healthcare: A Comprehensive Review and Directions for Future Research." *Applied Sciences (Switzerland)*. https://doi.org/10.3390/app9091736.

Khurshid, Anjum. 2020. "Applying Blockchain Technology to Address the Crisis of Trust During the COVID-19 Pandemic." *JMIR Medical Informatics* 8 (9): e20477. https://doi.org/10.2196/20477.

Kishan, K.V. Radha. 2007. "Structural Biology, Protein Conformations and Drug Designing." *Current Protein & Peptide Science*. Bentham Science Publisher. http://dx.doi.org/10.2174/138920307781369454.

Koonce, Lance. 2017. "Blockchains & Food Security in the Supply Chain." https://futurefoodtechnyc.com/wp-content/uploads/2017/05/Blockchains-and-Food-Security-in-the-Supply-Chain.pdf (accessed August 4, 2020).

Krittanawong, Chayakrit, Albert J. Rogers, Mehmet Aydar, Edward Choi, Kipp W. Johnson, Zhen Wang, and Sanjiv M. Narayan. 2020. "Integrating Blockchain Technology with Artificial Intelligence for Cardiovascular Medicine." *Nature Reviews. Cardiology* 17 (1): 1–3. https://doi.org/10.1038/s41569-019-0294-y.

Li, Lin, Lixin Qin, Zeguo Xu, Youbing Yin, Xin Wang, Bin Kong, Junjie Bai, et al. 2020. "Artificial Intelligence Distinguishes COVID-19 from Community Acquired Pneumonia on Chest CT." *Radiology* 296 (2): E65–E71

Li, Qun, Xuhua Guan, Peng Wu, Xiaoye Wang, Lei Zhou, Yeqing Tong, Ruiqi Ren, et al. 2020. "Early Transmission Dynamics in Wuhan, China, of Novel Coronavirus-Infected Pneumonia." *New England Journal of Medicine* 382 (13): 1199–207. https://doi.org/10.1056/NEJMoa2001316.

Liang, Xueqin, Zheng Yan, and Peng Zhang. 2016. "Security, Privacy, and Anonymity in Computation, Communication, and Storage." In *International Conference on Security, Privacy and Anonymity in Computation, Communication and Storage (SpaCCS 2016)*, no. December, 155–67. https://doi.org/10.1007/978-3-319-72395-2.

Livingston, Edward, Angel Desai, and Michael Berkwits. 2020. "Sourcing Personal Protective Equipment During the COVID-19 Pandemic." *JAMA*. https://doi.org/10.1001/jama.2020.5317.

Malsang, Isabel. 2020. "World Could Face Food Crisis in Wake of Coronavirus: UN, WTO." https://phys.org/news/2020-04-world-food-crisis-coronavirus-wto.html (accessed August 20, 2020).

Mambou, Sebastien Jean, Petra Maresova, Ondrej Krejcar, Ali Selamat, and Kamil Kuca. 2018. "Breast Cancer Detection Using Infrared Thermal Imaging and a Deep Learning Model." *Sensors (Switzerland)* 18 (9). https://doi.org/10.3390/s18092799.

Mao, Dianhui, Zhihao Hao, Fan Wang, and Haisheng Li. 2018. "Innovative Blockchain-Based Approach for Sustainable and Credible Environment in Food Trade: A Case Study in Shandong Province, China." *Sustainability (Switzerland)* 10 (9). https://doi.org/10.3390/su10093149.

Markov, Alexander. 2018. "Using Blockchain in Pharmaceuticals and Medicine." https://miningbitcoinguide.com/technology/blokchejn-v-meditsine (accessed August 4, 2020).

Mascaro, Lisa. 2020. "'We Have a Deal.' Federal Officials Reach Agreement on $2 Trillion Coronavirus Aid Package." https://time.com/5808667/congress-aid-package-disagreement/ (accessed August 14, 2020).

Murgia, Madhumita. 2020. "AI-Designed Drug to Enter Human Clinical Trial for First Time." https://www.ft.com/content/fe55190e-42bf-11ea-a43a-c4b328d9061c (accessed August 24, 2020).

Nero, Tracy L., Michael W. Parker, and Craig J. Morton. 2018. "Protein Structure and Computational Drug Discovery." *Biochemical Society Transactions* 46 (5): 1367–79. https://doi.org/10.1042/BST20180202.

Nichol, Peter B., Jeff Brandt, D. Kleinke, Angelo E. Volandes, Massachusetts General Hospital, Ross Koppel, Don E. Detmer, et al. 2016. "Co-Creation of Trust for Healthcare: The Cryptocitizen Framework for Interoperability with Blockchain." *ResearchGate* 24 (1): 1–9. https://doi.org/10.13140/RG.2.1.1545.4963.

Niller, Eric. 2020. "An AI Epidemiologist Sent the First Warnings of the Wuhan Virus." https://www.wired.com/story/ai-epidemiologist-wuhan-public-health-warnings/ (accessed August 13, 2020).

Omar, Abdullah, Shahriar Rahman, Anirban Basu, and Shinsaku Kiyomoto. 2017. "MediBchain: A Blockchain Based Privacy Preserving Platform for Healthcare Data." 534–43. https://doi.org/10.1007/978-3-319-72395-2_49.

Pajares, Gonzalo. 2015. "Overview and Current Status of Remote Sensing Applications Based on Unmanned Aerial Vehicles (UAVs)." *Photogrammetric Engineering and Remote Sensing* 81 (4): 281–329. https://doi.org/10.14358/PERS.81.4.281.

Perrella, A., N. Carannante, M. Berretta, M. Rinaldi, N. Maturo, and L. Rinaldi. 2020. "Editorial – Novel Coronavirus 2019 (SARS-CoV2): A Global Emergency That Needs New Approaches?" *European Review for Medical and Pharmacological Sciences* 24 (4): 2162–64. https://doi.org/10.26355/eurrev_202002_20396.

Reilly, Caitlin. 2020. "Blockchain Could Transform Supply Chains, Aid in COVID-19 Fight." https://www.rollcall.com/2020/03/31/blockchain-could-transform-supply-chains-aid-in-covid-19-fight/ (accessed April 23, 2020).

Reuters. 2020. "Expert Views: India's First-Quarter GDP Contracts by Record 23.9% y/Y" (accessed October 18, 2020).

Richardson, Peter, Ivan Griffin, Catherine Tucker, Dan Smith, Olly Oechsle, Anne Phelan, and Justin Stebbing. 2020. "Baricitinib as Potential Treatment for 2019-NCoV Acute Respiratory Disease." *The Lancet* 395 (10223): e30–31. https://doi.org/10.1016/S0140-6736(20)30304-4.

Rosser, James C., Vudatha Vignesh, Brent A. Terwilliger, and Brett C. Parker. 2018. "Surgical and Medical Applications of Drones: A Comprehensive Review." *JSLS: Journal of the Society of Laparoendoscopic Surgeons* 22 (3). https://doi.org/10.4293/JSLS.2018.00018.

Saleh, H, S. Avdoshin, and A. Dzhonov. 2019. "Platform for Tracking Donations of Charitable Foundations Based on Blockchain Technology." In *2019 Actual Problems of Systems and Software Engineering (APSSE)*, 182–87. https://doi.org/10.1109/APSSE47353.2019.00031.

Sang, Shengtian, Zhihao Yang, Lei Wang, Xiaoxia Liu, Hongfei Lin, and Jian Wang. 2018. "SemaTyP: A Knowledge Graph Based Literature Mining Method for Drug Discovery." *BMC Bioinformatics* 19 (1): 193. https://doi.org/10.1186/s12859-018-2167-5.

Scheraga, H.A., A. Liwo, S. Oldziej, C. Czaplewski, J. Pillardy, J. Lee, D.R. Ripoll, et al. 2006. "The Protein Folding Problem." *Lecture Notes in Computational Science and Engineering* 49: 90–100. https://doi.org/10.1146/annurev.biophys.37.092707.153558.

Schootman, M., E.J. Nelson, K. Werner, E. Shacham, M. Elliott, K. Ratnapradipa, M. Lian, and A. McVay. 2016. "Emerging Technologies to Measure Neighborhood Conditions in Public Health: Implications for Interventions and Next Steps." *International Journal of Health Geographics* 15 (1): 1–9. https://doi.org/10.1186/s12942-016-0050-z.

Schwartz, Eric Hal. 2020. "Microsoft Offers COVID-19 Chatbot to Help Healthcare Providers Triage Patients." https://voicebot.ai/2020/03/23/microsoft-offers-covid-19-chatbot-to-help-healthcare-providers-triage-patients/.

Senior, Andrew W, Richard Evans, John Jumper, James Kirkpatrick, Laurent Sifre, Tim Green, Chongli Qin, et al. 2020. "Improved Protein Structure Prediction Using Potentials from Deep Learning." *Nature* 577 (7792): 706–10. https://doi.org/10.1038/s41586-019-1923-7.

Shen, Bingqing, Jingzhi Guo, and Yilong Yang. 2019. "MedChain: Efficient Healthcare Data Sharing via Blockchain." *Applied Sciences (Switzerland)*. https://doi.org/10.3390/app9061207.

Shuai, Wang, Bo Kang, Jinlu Ma. 2020. "A Deep Learning Algorithm Using CT Images to Screen for Corona Virus Disease (COVID-19)." *MedRxiv*, 1–28.

Singh, Gari, and Jonathan Levi. 2020. "MiPasa Project and IBM Blockchain Team on Open Data Platform to Support Covid-19 Response." *IBM*. https://www.ibm.com/blogs/blockchain/2020/03/mipasa-project-and-ibm-blockchain-team-on-open-data-platform-to-support-covid-19-response/ (accessed April 4, 2020).

Smith, Kendra L, Isabel Ramos, and Kevin C. Desouza. 2015. "Economic Resilience and Crowdsourcing Platforms." *JISTEM Journal of Information Systems and Technology Management* 12: 595–626. http://www.scielo.br/scielo.php?script=sci_arttext&pid=S1807-17752015000300595&nrm=iso.

Stephen, Remya, and Aneena Alex. 2018. "A Review on BlockChain Security." *IOP Conference Series: Materials Science and Engineering* 396 (1). https://doi.org/10.1088/1757-899X/396/1/012030.

Sylim, Patrick, Fang Liu, Alvin Marcelo, and Paul Fontelo. 2018. "Blockchain Technology for Detecting Falsified and Substandard Drugs in Distribution: Pharmaceutical Supply Chain Intervention." *Journal of Medical Internet Research*. https://doi.org/10.2196/10163.

Talukder, Asoke K., Manish Chaitanya, David Arnold, and Kouichi Sakurai. 2018. "Proof of Disease: A Blockchain Consensus Protocol for Accurate Medical Decisions and Reducing the Disease Burden." In *Proceedings – 2018 IEEE SmartWorld, Ubiquitous Intelligence and Computing, Advanced and Trusted Computing, Scalable Computing and Communications, Cloud and Big Data Computing, Internet of People and Smart City Innovations, SmartWorld/UIC/ATC/ScalCom/CBDCo*, no. December, 257–62. https://doi.org/10.1109/SmartWorld.2018.00079.

Tann, Wesley Joon-Wie, Xing Jie Han, Sourav Sen Gupta, and Yew-Soon Ong. 2018. "Towards Safer Smart Contracts: A Sequence Learning Approach to Detecting Security Threats." http://arxiv.org/abs/1811.06632.

Thind, Gurpreet. 2020. "U.S. Adds Blockchain in COVID-19 Critical Services List." https://www.cryptopolitan.com/blockchain-in-covid-19-critical-services/ (accessed July 6, 2020).

Topol, Eric. 2020. "The U.S. Betrayed the Healthcare Workers Fighting the Coronavirus." https://www.fastcompany.com/90485838/the-u-s-betrayed-the-healthcare-workers-fighting-the-coronavirus (accessed May 4, 2020).

Tseng, Jen Hung, Yen Chih Liao, Bin Chong, and Shih Wei Liao. 2018. "Governance on the Drug Supply Chain via Gcoin Blockchain." *International Journal of Environmental Research and Public Health*. https://doi.org/10.3390/ijerph15061055.

United Nations SDSN Report, 2020. 2020. "Data and COVID-19." https://www.unsdsn.org/data-and-covid-19 (accessed April 4, 2020)

Wang, Juan, and Guoqiang Du. 2020. "COVID-19 May Transmit through Aerosol." *Irish Journal of Medical Science (1971–)*. https://doi.org/10.1007/s11845-020-02218-2.

Wang, Weiyu, and Keng Siau. 2018. "Trust in Health Chatbots." In *International Conference on Information Systems (ICIS)*, December.

Wang, Yunlu, Menghan Hu, Qingli Li, Xiao-Ping Zhang, Guangtao Zhai, and Nan Yao. 2020. "Abnormal Respiratory Patterns Classifier May Contribute to Large-Scale Screening of People Infected with COVID-19 in an Accurate and Unobtrusive Manner." http://arxiv.org/abs/2002.05534.

Weigend, Andreas S. 1994. *Time Series Prediction Forecasting The Future And Understanding The Past*. 1st ed. New York: Taylor & Francis. https://doi.org/10.4324/9780429492648.

White, Nick. 2020. "Global Coalition to Accelerate COVID-19 Clinical Research in Resource-Limited Settings." *The Lancet* 2 (20): 19–21. https://doi.org/10.1016/S0140-6736(20)30798-4.

Wiggers, Kyle. 2020. "Miso Robotics Deploys AI Screening Devices to Detect Signs of Fever at Restaurants." https://venturebeat.com/2020/03/24/miso-robotics-deploys-ai-screening-devices-to-detect-coronavirus-covid-19-fever (accessed April 2, 2020).

Williamson, Graham. 2020. "COVID-19 Epidemic Editorial." *The Open Nursing Journal* 14 (1): 37–38. https://doi.org/10.2174/1874434602014010037.

Yan, Li, Hai-Tao Zhang, Yang Xiao, Maolin Wang, Chuan Sun, Jing Liang, Shusheng Li, et al. 2020. "Prediction of Criticality in Patients with Severe Covid-19 Infection Using Three Clinical Features: A Machine Learning-Based Prognostic Model with Clinical Data in Wuhan." *MedRxiv*. https://doi.org/10.1101/2020.02.27.20028027.

Yu, Qing, Hui Liu, and Ning Xiao. 2018. "Unmanned Aerial Vehicles: Potential Tools for Use in Zoonosis Control." *Infectious Diseases of Poverty* 7 (1): 1–6. https://doi.org/10.1186/s40249-018-0430-7.

Yue, Xiao, Huiju Wang, Dawei Jin, Mingqiang Li, and Wei Jiang. 2016. "Healthcare Data Gateways: Found Healthcare Intelligence on Blockchain with Novel Privacy Risk Control." *Journal of Medical Systems* 40 (10): 218. https://doi.org/10.1007/s10916-016-0574-6.

Zeng, Tianyu, Yunong Zhang, Zhenyu Li, Xiao Liu, and Binbin Qiu. 2020. "Predictions of 2019-NCoV Transmission Ending via Comprehensive Methods." http://arxiv.org/abs/2002.04945.

Zhou, Lijing, Licheng Wang, and Yiru Sun. 2018. "MIStore: A Blockchain-Based Medical Insurance Storage System." *Journal of Medical Systems* 42 (8): 149. https://doi.org/10.1007/s10916-018-0996-4.

Zīle, Kaspars, and Renāte Strazdiņa. 2018. "Blockchain Use Cases and Their Feasibility." *Applied Computer Systems* 23 (1): 12–20. https://doi.org/10.2478/acss-2018-0002.

Zou, Lirong, Feng Ruan, Mingxing Huang, Lijun Liang, Huitao Huang, Zhongsi Hong, Jianxiang Yu, et al. 2020. "SARS-CoV-2 Viral Load in Upper Respiratory Specimens of Infected Patients." *New England Journal of Medicine* 382 (12): 1177–79. https://doi.org/10.1056/NEJMc2001737.

6 Machine Learning-Based Text Mining in Social Media for COVID-19

Tajinder Singh and Madhu Kumari

CONTENTS

6.1	Introduction	95
	6.1.1 Activities Involved in COVID-19 Text Mining Process	97
6.2	Value of Text Mining and Motivation	98
	6.2.1 Applications of Text Mining	99
	6.2.2 Design Issues in Text Mining	102
6.3	General Outline of Text Pre-Processing in COVID-19	103
	6.3.1 Main Challenges Related to Social Media Text of COVID-19	103
	6.3.2 Text Pre-Processing	105
	6.3.3 Scope and Recent Pre-Processing Approaches in COVID-19	107
6.4	Various Text Pre-Processing Approaches for COVID-19	107
6.5	Extraction Mechanism and Analysis of Social Text of COVID-19	110
6.6	Scope of Various Machine Learning Approaches in Text Mining in Social Media for COVID-19	112
6.7	Recent Approaches for COVID-19 and Their Scope	113
6.8	Prediction and Analysis of the Impact of COVID-19 on Different Parameters	118
6.9	Conclusion	119
References		120

6.1 INTRODUCTION

In the current era, social media is a unique primary asset for textual corpus and real-time social text streams. Numerous varieties of HTML documentations include vast and deep information, attracting a variety of users. Various types of social media users post both essential and redundant information on social media to seek the attention of other users. In the COVID-19 pandemic, social media is playing a key role and has become prominent in multiple ways which deal not only with text but also with multimedia. Information posted by a variety of users can be extracted over time using analysis tools which help to understand the real-time situation with changes over time. Similar to data mining, text mining looks for suitable information from the collected social streams or corpus and identifies interesting patterns. Every social media is jam-packed with vast data concerned with COVID-19, and such platforms are a great source for information-seekers who collect data and identify valuable patterns using analysis. Such types of social media text are usually extracted in an unstructured form which needs to be tamed for further analysis and classification purposes. Online social media is a source of information-sharing which provides extensive ways to contribute to and participate in online communities to become a part of an information diffusion system. Several kinds of social media have expanded and attracted great attention of social media users upto peak level. This availability of social media users and data in real-time situations helps to extract and analyze the information from a variety of perspectives. Numerous methods and approaches are given by various authors to extract social media text. These methods are used to distribute and connect with other online users to exchange information in multiple forms (Lifna and Vijayalakshmi, 2015).

Such kinds of platforms are contributing a lot to producing and spreading social information in real time and for seeking the attention of other users, by which the volume of social media users are exponentially emerging. With the rapid evolution of COVID-19, the World Wide Web (WWW) is fully occupied with news and articles related to COVID-19. The WWW has become a leading source for addressing the concerns of social media users and for developing eminent data for real-time social stream data, including a large amount of text corpuses. In one study (Goldhahn et al., 2012), numerous schemes which help to assemble textual information concerned with COVID-19 are explained. The assembled text needs to be processed before further analysis. Therefore, taming the text is a valuable step through which the necessary and highest quality information can be pulled out. The extracted data patterns can be used for multiple purposes such as polarity disambiguation and sentiment analysis (Bhadane et al., 2015; Hamdan et al., 2015; Saleiro et al., 2017), social event detection, classification, and analysis on social media (Vavliakis et al., 2013; Zhou and Chen, 2014; Zhao and Mitra, 2007).

Nowadays, as COVID-19 is gaining more attention, it is also forcing people to change their lifestyles and social life. This trend of change is also a key aspect where the recommendations advisory is passed by various health organizations through online and offline modes. By including all series of patterns related to COVID-19, it is observed that with the change of time, its trend keeps on growing, and various recommendations are passed in terms of precautions. So it's not hard to say that trend prediction (Aiello et al., 2013), topic tracking, and recommendation (Lin et al., 2011; Martínez et al., 2008) of the COVID-19 pandemic are related to each other either directly or indirectly. Multiple methods are used by social media users to represent information on social media, and for textual information, the (Bag of Words) BOW model has become very common due to its simplicity. In this COVID-19 situation, the main goal is to understand the type of information shared on social media, in which name entity recognition is playing a significant role (Quan and Ren, 2014). It helps to understand the nature of the collected text corpus or text stream to mine useful patterns of information (Croft et al., 2010) concerned with COVID-19.

Text Mining in Social Media

To bring more clarity to this vast area of text mining, its key roles in the COVID-19 pandemic are explained below:

- *Information exploration and retrieval*: Text related to COVID-19 is to be explored and retrieved by a user who wants to understand the hidden features of the extracted data. The extraction of data can be based on keywords, where a user will pass a query to extract information, or indexed-based (Croft et al., 2010; Alves et al., 2019).
- *Clustering of extracted text*: To collect the text related to COVID-19 from the extracted data or in the real-time scenario, clustering plays a key role (Ghai et al., 2016). Multiple algorithms are available to cluster the data, whereas online fast clustering is effectively used for fast processing of data in real time (Aggarwal, 2014).
- *Classification of collected text*: In text mining, classification is an important task which classifies the collected text into various segments. In the COVID-19 text corpus, if a user wants to extract only related information then classification comes into the real picture (Santana et al., 2020).
- *Mining of web*: COVID-19 related information is available in huge amounts on the web in distinctive structures. Web crawlers are used to extract the data, usually in a structured way from which patterns of information can be analyzed by applying machine learning algorithms (Desikan et al., 2006).
- *Extraction of features and concept*: Features from the collected text are extracted and the collected text usually is unstructured. The attributes and features of collected data can be visualized by extracting the features using machine learning algorithms. Similarly, if linked information is existing in the collected text then the concept of linkage is also analyzed and matched with the available text (Popowich, 2005; Gelfand et al., 1998).

6.1.1 ACTIVITIES INVOLVED IN COVID-19 TEXT MINING PROCESS

Activity 1: Create the corpus: This is the first activity involved in the text mining process. The main motive of this activity is to develop a corpus from the collected text in an appropriate way. In this step, all the features are combined together, which helps to analyze the hidden patterns accurately (Figure 6.1). In this step, COVID-19 related features are obtained and are analyzed further for a detailed study.

Activity 2: Text pre-processing: This activity removes unwanted data from the collected text. Structured representation with a detailed attribute value comes out after text pre-processing. The main aim of this activity is to remove slang, emoticons, abbreviations, misspelled words (Lifna and Vijayalakshmi, 2015; Kumar and Govindarajulu, 2013), and other related information which is not required for the study of the COVID-19 text.

Activity 3: Knowledge extraction: New, well-structured, knowledgeable information is extracted from the processed text in the situation of exact dilemma being addressed using the structured COVID-19 text. Various activities are part of this activity, such as prediction, which involves classification and real-time series

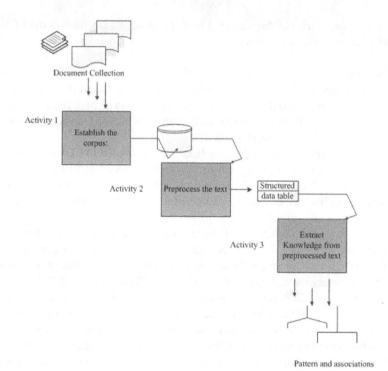

FIGURE 6.1 Representation of COVID-19 text mining task.

prediction, association and clustering of texts, trend prediction, and analysis to extract informative patterns.

For this purpose, this chapter is divided into various sections. Section 6.2 highlights the value of text mining and motivation, and Section 6.3 elaborates the general outline of text pre-processing. Similarly, Section 6.4 explains the various text pre-processing mechanisms helpful in COVID-19, and the extraction mechanism of social texts is described in Section 6.5. The scope of various machine learning approaches in text mining in social media for COVID-19 and recent approaches of text mining are explained in Sections 6.6 and 6.7, respectively. Impacts of the COVID-19 pandemic in various sectors are explained in Section 6.8, and Section 6.9 puts forward a conclusion.

6.2 VALUE OF TEXT MINING AND MOTIVATION

Text mining in the current era has become an essential element of social media usage. In this COVID-19 situation, social media is in huge demand and is attracting increasing numbers of users on a regular basis. Every day numerous users join social media platforms and post their sentiments and emotions related to COVID-19. Machine learning and its multimodality potential allow social media users to extract and collect informative patterns from the posted text. Multiple machine learning approaches can play a key role in this pandemic in detecting, classifying, and analyzing features of COVID-19 in real time. These features are arranged in such a way that the machine learning mechanism can understand them effectively at superior granularity. We have seen that the traditional text mining mechanism is not efficient at dealing with higher granularity, thus leading to lower efficiency. But due to the smart, intelligent, and efficient behavior of advances, the machine learning mechanism provides a platform to perform complex tasks in an easy way in real time. During COVID-19, social media users' texts are not tamed as they contain unwanted information and they must be processed before further analysis. It is also observed that ambiguity is a major concern in social media and can be a problem in analyzing a text. For this reason, domain knowledge and actual features extraction require an advanced learning approach which iscapable of handling missing values, unwanted information, abbreviations, shortcuts, and contextual information as well. Thus, in this chapter we explain various machine learning approaches for text mining in social media. Numerous approaches for extracting semantic and contextual information which provides empirical properties of social texts and social interactions of users are also explained. Due to the significant role of text mining in many domains, we are particularizing it more on COVID-19.

6.2.1 Applications of Text Mining

Applications of text mining are available in numerous areas, and if we classify them according to the goals, then it can be categorized as defined below:

- *Review and open-ended evaluation*: Seeking customers' attitudes and analyzing their behaviors in market text mining are used very well. In marketing, user reviews and sentiments are an integral part of analysis, whereas open-ended surveys are also used for exploring customers' attitudes. The main motive of this kind of evaluation is to allow customers to articulate their analysis in terms of sentiments.
- *E-mail, online posts, and text message processing*: In text mining, classification is a primary task of text in real time or offline mode. Classification of online text messages is a common practice to tackle an unwanted situation in terms of receiving spam emails, and spam messages are to be filtered out. In this way, the online traffic which is not desired can be controlled/discarded.
- *Evaluating warranty, insurance entitlements, and analytical interviews*: In a digital business environment, a huge amount of information is collected using various modes of information collection. Electronically, as we have seen in the COVID-19 situation, brief interviews can be conducted with medical experts and healthcare teams to learn about the symptoms and methods to be safe. Such kind of textual keywords in various scenarios can be used to extract informative patterns after processing.
- *Finding and analyzing competition*: Extracting text from a website helps to analyze the specific area automatically to determine the nature of users

of a particular website. It helps to process the web contents and pages to seek various documents. These available documents can be used to extract informative patterns which help to determine multiple features.
- *Semantics mapping*: Contextual information has a great essence in a social media text. As users upload posts on social media related to COVID-19 which are directly or indirectly associated with each other contextually, therefore, to understand the relation among posts and their related context, machine learning plays a significant role.
- *Web searching*: In web exploration, meta search engines and their great potential to respond to the user's queries are gaining great attention. These search engines enhance web exploration, where one search engine sends out a search request to other search engines and displays all the results. In this way, the significance and outcome of browser and browsing session can be enhanced, and it is possible with this meta search engine only.
- *Classification and analysis of sentiments*: The area of sentiment analysis and classification in COVID-19 is helping researchers to find opinions of the public related to this pandemic. In real-time sentiment analysis, text can be classified into different classes such as positive, negative, or neutral. But in this area, polarity disambiguation is a big issue at text and document levels. The work of sentiment analysis is to categorize social media users' text to understand their actual experience and behavior.
- *Mining bibliographic data*: Text mining and machine learning applications play a key role in extracting data related to references. Mining bibliographies and extracting duplicate information help to arrange a relationship between various authors. It also helps to analyze and visualize the functioning of bibliographic text.
- *Heated language detection and cybercrime*: It is observed that in social media, often users send anonymous messages which include abusive words. Therefore, to analyze the root node for this purpose, a precise approach must be developed which detects heated speech and antagonistic words from social media groups or blog posts. Machine learning approaches are contributing very well in this area to extract chat words from various chat platforms and detect heated speech and cybercrime.
- *Social media event detection, classification and trend analysis*: Numerous events related to COVID-19 are occurring on social media platforms. Social media users participate in these events and give their own views. It's a good opportunity to analyze the nature of events and to classify them according to their features. Sometimes it is also good to divide the events into different categories to analyze the trend and their social impact. Machine learning approaches are becoming increasingly advanced and can predict and analyze the trends and social impact of a particular event such as COVID-19. Novelty detection is also another feature, which captures features and decides their novelty accordingly.
- *Spam detection in social profiles*: Often users post information on social media using a fake profile. The information diffused from such profiles

will usually be a rumor or fake, whose main motive is to spread an advertisement through a URL. In this COVID-19 pandemic, we have seen that rumors are also diffused in various forms like videos, text, and posts on blogs. Classification approaches are used to classify spam profiles, links, and URLs, which block the spammers from distributing spam messages and promoting personal blogs, advertisements, and pages, as described in Figure 6.2 clearly.

Now we have seen that the area of machine learning-based text mining is used for a wide array of applications spanning multiple domains. We can say that numerous approaches, methods, techniques, and tools are available which can help to extract information from various social media platforms.

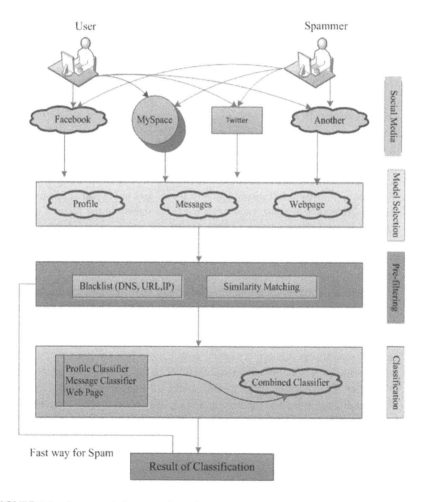

FIGURE 6.2 Framework for spam detection source.

6.2.2 Design Issues in Text Mining

Usually text on social media is noisy and unstructured. Due to this unstructured nature, it must be processed before analysis. In COVID-19, as we have analyzed the nature of the text, it is observed that the attributes of the extracted data are not clear. This unclarity of data leads us to process the collected data to extract knowledgeable patterns to take a decision. Therefore, to clean text and to find the knowledgeable patterns from collected text pre-processing plays a key role and comes into real picture. During the extraction process, we also observed that social media users use abbreviations, slang, short forms, and emoticons, which are very difficult to understand. Therefore, in the COVID-19 pandemic, every social media user is typing on social media with multiple syntaxes, short forms, slangs, abbreviations, and semantics which are used to define the importance of specific words.

Due to this reason, contextual information is also very important to represent the genuine meaning of textual words on social media in sentences and documents. Multiple issues and designing strategies are available to design and study the COVID-19 methodology in text mining and are described below:

- *Multimodality*: Texts on social media contain information in the form of multimedia, which is usually a combination of images, text, URLs, tags, and hyperlinks. In COVID-19, we are dealing with text only and we are not processing other forms of information. Therefore, it is essential to process the collected text for analysis, and for this purpose, machine learning approaches are used.

  The unwanted information which is not required is to be deleted (Singh and Kumari, 2016), otherwise it leads to social text noise and ambiguity. In the text extraction process, we extract the information in the form of social streams in real time related to COVID-19 (Bhadane et al., 2015; Bhuta et al., 2014). The extracted information can be used for classification, trend prediction, and analysis of the impact of COVID-19 on social life and other aspects.
- *Rich social context*: Rich contextual information is available in social text streams, as in the real-time extraction numerous social media users interact with each other from different platforms and exchange their ideas and views. This form of social connection of users in the form of information, generates an independent connection with each other. In these social connections nodes are connected with each other in the form of tree, graph, and in another means which helps to extract the all features together. But it is also a challenging task to identify and study the features which are hidden in collected text and to classify it on the basis of contextual information as well. In text mining, parsing is used to deal with rich textual information to analyze the hidden features to extract informative patterns.
- *Inconsistent quality*: The social text is unstructured in nature, which means that it is a collection of all types of data, including text, images, URLs, hyperlinks, emoticons, slang, and abbreviations. The presence of such

things degrades the quality of the text, or in other words, we can say they are noise in social texts. Therefore, the collected data streams of COVID-19 are pre-processed to reduce the impact of such noise on the resultant information.
- *Huge volume and multisource*: Multiple sources are platforms of social media that are available, where numerous users are posting their views by multiple means. The information shared by users on social media can go viral within few minutes, or it may take time to grab the attention of other users. It depends on the content and quality of the information and active social participation of social media users. In COVID-19, we collected text in real time in the form of social texts, which is a big issue in text mining to handle text generated from numerous sources which are huge in capacity and in providing significant analysis related to an event such as COVID-19.
- *Dynamic in nature*: Social media text in a real-time scenario is dynamic in nature as it changes with time. As COVID-19 cases are increasing rapidly, social media is getting multiple posts every few seconds. To capture the information and to analyze the features of real-time data, clustering plays an important role. Usually fast clustering is used, which collects similar text at the same time effectively and efficiently. Therefore, to maintain quality/correctness of the collected data and to prevent wastage of resources because of fast updates, the text should be identified and handled in a timely manner in the real-time situation (COVID-19).

It is very clear from the above discussion that in order to design a methodology there are many challenges, and to sustain the correctness with efficiency identification of textual features with regard to their word sense and recovery is important. Due to isolated conditions, the text process should have capabilities to cope up with such correctness without any human intervention.

6.3 GENERAL OUTLINE OF TEXT PRE-PROCESSING IN COVID-19

Text pre-processing contains a number of steps in a sequence designed to convert noisy text into an appropriate form for input into an algorithm. In Figure 6.3, the distinctive operations of text pre-processing for COVID-19 analysis are illustrated. Tokenization is the most important step because it represents COVID-19 information sparsely and unusually. In tokenization, the input text is divided into small units and the subsequent tokens are tagged with their respective part of speech (POS) and COVID-19 event diffusion rate. In the next phase, tokens are transformed into a reliable case using lemmatization, and finally, filtering of stop words is typically implemented.

6.3.1 MAIN CHALLENGES RELATED TO SOCIAL MEDIA TEXT OF COVID-19

COVID-19 in social media is an ambiguous notion which refers to generating and sharing of COVID-19 data or contributing to social networking (Peetz, 2015).

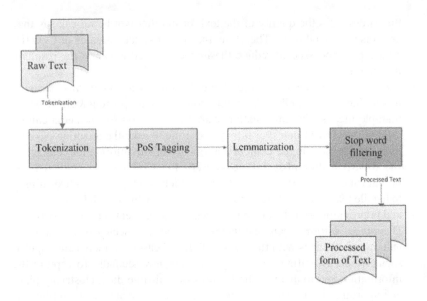

FIGURE 6.3 Basic layout of text mining.

Multiple blogs and social media platforms which are related to COVID-19 are available, and these all are contributing in extracting data and seeking knowledgeable patterns from it with great potential. Various methods such as question answering, sentiments, mailing lists, and group discussion platforms are designed to present information delicately to facilitate the people and provide precautions related to this pandemic (Ren, 2016; Han, 2014). In other words, we can say that in the current era, social media is including various tools and methods which allow users to share, access, and diffuse information. As social media is an integral part of users' experience, then we can say that social networking has become a feature of social media. Various authors such as Aichner and Jacob (2015) have divided social media into the different ways social media is represented, including microblogs, e-commerce portals, multimedia sharing, social networks, review platforms, social gaming, and virtual worlds. These domains of social media are very interesting to capture knowledge in terms of features associated with them. Texts of social media contain special features which are represented in Figure 6.3.

- *Tininess*: In social media, the size of users' posts is usually small when compared to traditional media. If we discuss on Twitter, then the user can post only 280 characters, as their posts, called "tweets," are limited to 280 characters (Ren et al., 2013). Due to the shortness of these tweets, it is very challenging and difficult to extract the required features related to COVID-19 from such a small length of text.
- *Multilanguage*: Users of social media can express their views using different languages and can choose any platform to discuss and they share their views in groups, blogs, communities, or at any social media platform.

TABLE 6.1
Social Text Quality Challenges

Challenge	Description
Stop list	Occurrence of frequent words in text
Text pre-processing	Remove the undesired information from collected text
Intelligibility of words	To provide a clear meaning in text
Tagging	Predicting data annotation and its characteristics
Syntax/grammar	Scope of ambiguity, data dependency
Tokenization	Various methods to tokenize words or phrases
Usual learning	Similarity measures and use of characterization

Languages used by the participants can be different from those of others and understanding the real meaning of multilingual texts on social media is becoming challenging. In the pandemic of COVID-19, if we are not able to keep up with the challenge of translation then it will become a problem for normal users to understand the actual system (Ren, 2016). Therefore, a machine learning mechanism is required to hande such challenges and provide clear-cut information to the people.

- *Opinions*: On social media, every post holds an opinion. Opinion can be defined as a quintuple (oj, fjk, soijkl, hi, tl), where, "oj" is a target object, "fjk" is a feature of the object "oj," and "soijkl" is the sentiment value of the opinion of the opinion holder "hi" on feature "fjk" of object "oj" at time "t." An opinion can be positive, negative, or neutral, which depends on the words expressed by the opinion holder (Pang and Lee, 2008).
- *Appropriateness*: The dynamic nature of social media provides precise information but its exact representation of textual documents varies with the change of time and sometimes the whole structure of the sentence also changes, including its meaning. Therefore, to analyze the exact picture of COVID-19 detection, classification, and diffusion prediction in the current scenario is quite challenging due to its dynamic behavior (Table 6.1).

6.3.2 TEXT PRE-PROCESSING

Text pre-processing is an important phase after the extraction of text from social media. In social media, as we have explained that text will be unstructured, it will include unwanted information, which is referred to as noise in text mining. Various text pre-processing methodologies are available to tame the social text. These methodologies give good performance and help users to analyze informative patterns. We know that in social media, users use short forms and abbreviations to communicate with each other. Few short forms are standardized, whereas many are user-specific and either need to be included in a dictionary or we have to tame them as per their exact meaning. If we consider the ISBN number of a book then it's a standard representation which refers to the "International Standard Book Number" (Cucerzan and Brill,

2004). But on social media and on other social platforms, users often use short forms or non-standard words, such as hru/how are you, ttyl/talk to you later, and cu/see you (Han, 2014). Sometimes, there are typos, which are referred to as non-standard words in social text mining and need to be converted into actual words (e.g., helo/hello, stndrd/standard), phonetic approximations (e.g., w8/wait, f9/fine), words with repetitions (e.g., hiiiiiiii/hi, hellooooooo/hello), informal abbreviations (e.g., awsummmmmm/ awesome, gr888888888/great). These non-standard words are called noise in social media. Such non-standard words are hard to analyze, and they make it complex and challenging to compute the exact meaning of words. Similarly, during the extraction of tweets related to COVID-19, similar challenges exist which demand to be addressed.

- *Influence of non-standard words*: Non-standard words on social media text degrade the quality and performance of the text. As we have extracted social text stream related to COVID-19 from Twitter, we observed that it contains numerous non-standard words. On Twitter, the availability of the tool to count the performance in terms of precision is not good (Gimpel et al., 2010). These tools are also not working well for part-of-speech tagging (Gimpel et al., 2010), syntactic dependency parsing (Cucerzan and Brill, 2004) and name entity recognition (Liu et al., 2011), and machine translation (Owoputi et al., 2013). Penn treebank is the best solution for identifying the part of speech, which helps to accomplish acquiring the required information, including tagging as well. Probability distribution and conditional probability are also good solutions (Marcus et al., 1993) which help to analyze non-standard words (Toutanova et al., 2003). Tagging of non-standard words should be performed accurately, otherwise the accuracy of the system will degrade with wrong tagging mechanism. Because the tagging mechanism is directly attached with the words collected, if it is done correctly, then it will give good accuracy, otherwise it will affect the output also (Klein and Manning, 2003). Various researchers have used the parsing mechanism for different purposes such as investigation of forum text (Foster, 2010) and text foundation diversity (Baldwin et al., 2013), which computes that the major reason which influences error in parsing is non-standard words which are triggered and influenced by wrong POS tagging output (Han, 2014). Consider an example in which non-standard English expressions *hv COVID-19 symptoms 2gther* is divided into word level and then the phases are normalized into the actual word form.
Before normalization: *dey unluckily 2 hv COVID-19 symptoms 2gther.*
After normalization: *They unluckily to have COVID-19 symptom together.*

In text mining, to represent a name entity, capital letters are used but sometimes the capital letters are used to represent an abbreviation on Twitter also (Ritter et al., 2011). This non-standard and informal nature of text disturbs performance in terms of accuracy (Pomares Quimbaya et al., 2016). Therefore, we can say that abbreviations, short forms, slang, and non-standard words are an integral part of social text (Singh and Kumari, 2016). In the classification task of COVID-19, the @ tag plays a key role in helping to extract the required information from a huge text.

6.3.3 SCOPE AND RECENT PRE-PROCESSING APPROACHES IN COVID-19

In the previous section we explained and discussed the influence of non-standard words on social media in terms of accuracy. Now in this section we will elaborate on the recent and most commonly used text pre-processing approaches. The main motive of these approaches is to assemble text in such a way that words can get their exact meaning after processing (Han et al., 2012; Sproat et al., 2001). We have seen that, in the case of COVID-19, contextual information plays a huge role in detecting text related to COVID-19 including events, classification, sentiment analysis, and ambiguity handling. These approaches are simple to use and help a user to find the contextual information of a particular keyword which helps in exact keyword counting (Daume and Marcu, 2006).

Normally social media words can be categorized into two classes: out-of-vocabulary (OOV) non-standard words and in-vocabulary (IV) non-standard words (Han, 2014). The pre-processing task becomes more complex when the words in the extracted text are OOV because such words require extra care to find their exact meaning. On the other hand, IV words have scope in pre-processing where their meaning and contextual information can be analyzed. The most common and widely used approach is the context insensitive lexical approach. This approach helps to find the exact meaning of words, and mostly it is used in those words where a single character exists multiple times such as goooood/good and hellllloooooo/hello.

On the other hand, ambiguous words in COVID-19 text create a problem in a few cases, as it becomes very difficult to identify the exact meaning and sense of a word. Sometimes in social texts if the exact meaning of the word is available but the contextual meaning is different, then the situation becomes inadequate. This can be explained by the following example:

Police administration charges rupees 10 thousand fine for not wearing mask in public

Whereas, in the second example:

I hope u will be fine in this epidemic.

In the above examples, fine represents both "money" and "fine" is also used to express human feelings and perspectives about self. Thus, in pre-processing to handle ambiguity, contextual information outside the actual non-standard word should be addressed. In the next section we explain the key role of existing text pre-processing approaches with their significant role in different domains.

6.4 VARIOUS TEXT PRE-PROCESSING APPROACHES FOR COVID-19

Machine learning translation-based text pre-processing for Covid-19: Machine learning translation is an advanced approach which is used for text pre-processing due to its good performance. We have extracted COVID-19 text stream from social media, which includes numerous types of data. The stream of collected text can be categorized as wanted and unwanted information. The features which are required for making a decision are standard features which are decided by users before processing and others are discarded. Multiple approaches are available which contribute very effectively in the area of text pre-processing in which phrase level (Aw et al., 2006; Koehn et al., 2007) and supervised machine learning-based translations are used very frequently. These machine learning approaches are based on keywords in which repeated characters existing in the keywords are removed to obtain the actual word to analyze the impact before translation (Kaufmann and Kalita, 2010). In this way, accuracy and effectiveness of various machine learning approaches can be measured and, on the other hand, statistical machine learning helps to analyze the feature of texts concerned with online social media users. Features are generated automatically and analyzed by machine learning approaches based on the training set, and in this way, language models for translations are used to evaluate their performance also (Schlippe et al., 2010). In machine learning-based text pre-processing, if we want to analyze the COVID-19 text based on a single character then character-based translation comes into the picture. It helps to analyze the information in a combined way including all features which exist in a text. Secondly, character-based machine learning translation is less sparse, due to which text can be processed including all features into their normal standard form (Pennell and Liu, 2011).

- *Text pre-processing based on the spell-checking approach for crawled COVID-19 text*: Informal and slang-based communication on social platforms is very common, whereby social media users use short forms and abbreviations for many words. Such kinds of situations create a problem to analyze the text, which also becomes dangerous when it gets mixed with typos. It becomes a critical situation for the researcher when the related keywords cannot be clearly classified. It is also observed that typos that unintentionally occur by typing incorrect spelling, such as "fear" to "fair" or "peace" to "piece," can be solved by analyzing the contextual information and meaning. But if the context is not clear then it will be a challenge to find the contextual meaning to analyze the actual sense of the keywords existing in a sentence. Around 40% of typos occur unintentionally in effective keywords (Kukich, 1992), and in this COVID-19 pandemic, social media is overloaded with various kinds of information and it is essential to study and analyze the keyword spelling to identify their impact. Because if the keyword holds the useful information then it will be good to analyze, otherwise it can be ignored (Whitelaw et al., 2009; Wilcox-O'Hearn, 2014). UNIX-based dictionary can be used to correct words. Ahmad and Kondrak (2005) used a similar dictionary approach to correct non-standard words into actual standard words. Therefore, the classification and analysis of standard and non-standard words in text mining is very important. In the COVID-19 situation, the main motive is to avoid rumor-based information and derive a mechanism which will analyze the users associated on the social media platform and identify their profile. Fake profile users process and post rumors on social media and later fake information can become viral due to active participation of social media users. Various events on social media happen and the events related to COVID-19 are occurring in a huge amount these days. Due to this high availability, classification and analysis of related events can be analyzed.

Numerous approaches such as n-gram, uni-gram, and Google n-gram can be used to analyze the text. Spell-checking and finding keyword meaning are very common and can be performed to improve the performance of the approach. But if similar keywords exist in the collected text then the distribution of various feature words is to be evaluated to understand the correct word meaning (Lin, 1998). Because it is observed that when we extract text from social media, numerous words are referred to as non-standard words as their meaning is not available (Li et al., 2006). In social media, words like f9, b4, str8, and many more are OOV and will not be recognized by spellchecker, and extra efforts are required to analyze such words. To understand and analyze the behavior of non-standard keywords, Cook and Stevenson (2009) have explained the unsupervised machine learning approach to perform pre-processing, where non-standard words are converted into standard words. This proposed approach also works for slang, and a similar kind of methodology is used by Emadi and Rahgozar (2020). Various aspects are considered to analyze the data features, of which four are important to explain: orthographic factor, phonetic factor, contextual factor, and acronym expansion. Information related to these four aspects can be directly extracted and analyzed for analysis and if we consider COVID-19, Twitter is a major source for extracting information. Using machine learning approaches, abbreviations such as "cu"/"see you" can be easily recognized.

- *Current pre-processing methodologies*: Various text pre-processing techniques exist as we have explained in the previous sections. These methodologies can be used for the pre-processing of COVID-19 extracted text, which can convert non-standard keywords into an understandable form. CRF (Conditional Random Fields) are also used in text pre-processing, and Chrupała (2014) used the CRF-based approach for text cleaning. Similarly, Bayesian-based text pre-processing methods can be used to clean the text. The non-standard words can be converted into meaningful words, and Bayesian-based approaches are helpful in this activity. These are also used for detection and classification of COVID-19 related keywords which refer to events in social media. Linear-based CRF are also used for text processing. Usually in text streams, typos and cognitive errors can be resolved by CRF nicely without compromising the performance of a system (Choudhury et al., 2007). Recurrent neural networks and hidden Markov models are also used to address such errors which capture the morphological features and process further to classify the text according to the features. On the other hand, unsupervised machine learning approaches are also used in text pre-processing, whereas Hassan and Menezes (2013) proposed and discussed a method which is based on unsupervised machine learning. Features of collected text are evaluated and represented as a graph, and a random walk mechanism is used to identify the contextual information. This will analyze the contextual meaning of text and a bipartite graph is generated using n-gram. The graph can be represented as $G(w,c,e)$, in which "w" represents whole sequence of words (collected noisy text), "c" represents shared contextual information for all nodes, and "e" represents edges in the graph. Edges in the graph join nodes containing words and contextual nodes. Weight updating mechanism is implemented, which calculates and updates the weight every time a contextual word arrives from the collected text. Therefore, such types of mechanisms can be used to analyze the COVID-19 textual information, and the contextual information can also be analyzed by extracting the correct meaning of the featured keyword.

In COVID-19 text stream analysis, if we implement the parsing mechanism for analysis then it will be a successful attempt to gain informative patterns. In many research articles, the parsing mechanism is implemented for various tasks such as pre-processing (Zhang et al., 2013). Therefore, in the pandemic, to identify the actual consequences of the collected non-standard words, a parser can be implemented directly. In the existing research, authors evaluated a parser's performance by comparing it with the existing methodologies, considering various metrics such as F-measure, precision, and recall (Yang and Eisenstein, 2013; Derczynski et al., 2015). Sarker (2017) designed an approach based on unsupervised machine learning for the next cleaning. The log linear method is used to analyze the standard and non-standard words, whose aim is to convert non-standard words to standard words. Translation decoder is used by Wang and Ng (2013), which also deals with missing values from the dataset. The decoder works in two phases: in the first phase, they deal with the missing values, and in the second phase, they also generate the hypothesis for the collected text.

6.5 EXTRACTION MECHANISM AND ANALYSIS OF SOCIAL TEXT OF COVID-19

- *Data collection*: In the COVID-19 study, extracted tweets from the Twitter API or from the corpus of social text related to COVID-19 are studied. For this chapter, we collect the tweets related to COVID-19 and accordingly keywords and hashtags are applied at different time slots of streaming text for extracting data. The extraction process was divided into various time slots to collect an average amount of tweets according to the character of the aimed event. With the use of time stamp information in the text stream, the dataset can be converted into stream. In this stream of tweets, each object contains text as well as the sender and receiver information including network structure.
- *Mining the ground reality*: In every text corpus or stream of text, a huge number of important matters are concealed. Therefore, to understand the reality of the text, an awe-inspiring amount of effort would be entailed to

extract features manually. In other words, we can say that we cannot rely totally on the extracted features from stream or the features visible in datasets. For each and every topic, the ground reality contains a multiple set of keywords in a particular time slot, in which features and theme appear in conventional reports, because in the COVID-19 case, the data is increasing very rapidly and the situation is also changing with time, including symptoms.

- *Theme detection*: Every text stream should be based on a theme which is to be evaluated on each time slot. Every stream of text is divided into various time slots and we merely consider texts from streams as input with ground reality.
- *Evaluation of theme detection output with ground reality*: Self-detection of a theme in streams of texts is evaluated by analyzing their behavior and its nature with regard to time. Therefore, machine learning approaches can be used, which can analyze the rapid change over time in the actual conduct of COVID-19 features and their ground reality, which can be trained by using various machine learning approaches.

From Figure 6.4, it is very clear that the data collection mechanism can be started many days before the finding of useful patterns of COVID-19 related events. The peaks in Figure 6.4 indicate the maximum traffic of tweets during extraction.

The COVID-19 topics of events are considered by various durations which are divided into different time slots. On the x-axis the time is represented in hours, and on the y-axis the number of COVID-19 related tweets is defined in thousands. Similarly, the utilization of memory is described in Figure 6.5 and it is also consistent. Java virtual machine (JVM) is used for system organization. In the whole scenario, the process of memory management is consistent but it is observed that

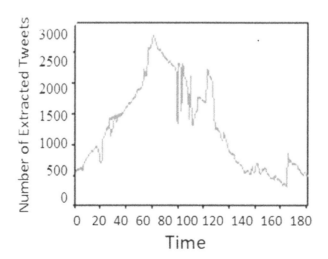

FIGURE 6.4 Process of data extraction for COVID-19.

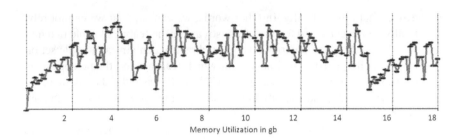

FIGURE 6.5 Memory utilization during process of COVID-19 data extraction.

in the beginning, there is a decrease in memory utilization due to less detection of COVID-19 related events, and the second reason behind less utilization is that the system needs time to stabilize long-lived real-time objects.

6.6 SCOPE OF VARIOUS MACHINE LEARNING APPROACHES IN TEXT MINING IN SOCIAL MEDIA FOR COVID-19

Nowadays, COVID-19 is a hot topic all over the world. Events related to COVID-19 are becoming a famous research area, which deals with the study of opinions, attitude, evaluation, and emotions in the form of entities such as place, location, time, symptoms, diffusion rate, and its social impact on users. All types of social platforms are sharing information related to COVID-19, and many social media users talk about it by sharing their emotions, sentiments, and personal views. To understand social media users' behavior and impact on the system, it is necessary to classify their posts into various classes such as positive, negative, or neutral. These classes are directly related to the words expressed by social media users, including contextual information. Time, location, and user profile matter a lot in sentiment analysis because a fake user can also write positive or negative on social platforms, but it is very hard to identify when and from where the sentiment is expressed, including the contextual and polarity information (Liu, 2010, 2012). We can say that sentiment and opinion analysis are similar areas but a mysterious difference exists between them (Liu, 2010, 2012). Sentiments are usually classified into positive and negative classes, whereas an opinion can be unpolarized. We can divide sentiment analysis mainly into three levels on the basis of COVID-19 text extraction, which are:

- *Document level*: To classify the collected textual documents in positive and negative classes is a very common and easy task which distributes the sentimental words into their classification class (Pang et al., 2002). Consider an example of a symptom, whether the patient shows COVID-19 symptoms or is asymptomatic. So it will be very difficult to analyze the patient, but when the data will be collected and analyzed, results will come out. Therefore, on the basis of the extracted sample, the result report is prepared to analyze the nature of a patient's current situation followed by history also. Compound

entities can also exist, which usually occurs to express their combined impact on the document level sentiment analysis.
- *Sentence level*: This deals with subjectivity classification (Wei and Pal, 2010), which expresses the sentiments into different classes as positive, negative, and neutral. Subjective views related to COVID-19 can be analyzed, and it is also observed how many times a similar sentiment/view is expressed by social media users, and accordingly weightage is to be given to a particular sentiment word. Sarcastic words and heated speeches are also analyzed in COVID-19 text streams, though it is a challenging task to analyze their impact on social media users and society. Consider an example such as "what a strong immunity! It stopped fighting against COVID-19 in two days." So if we analyze this example, then we can say sarcasm is not common in existing words, but in political blogs and posts it is very frequent.
- *Entity and aspect level*: In entity level and aspect-based analysis, COVID-19 features can be represented in different ways because we know that COVID-19 is referred to in many ways, such as CORONA, COVID-19, and SARS-CoV-2, which are associated with a single entity. Decent information and detailed analysis can be gained by representing a single entity in multiple ways (Hu and Liu, 2004). At aspect level, sentiment class can be directly recognized according to the target feature. Target-based analysis is a good approach to analyze the nature of sentiment as the predefined class of target helps to analyze the features from upcoming text. If we consider the example, "The country is full of good resources but the life is not secure," in the first stage of analysis it will give positive sentiment as it is expressing positive words about the country but negative words about security of life. Therefore, it is necessary to analyze such sentences as per their aspects and entities to gain full information.

6.7 RECENT APPROACHES FOR COVID-19 AND THEIR SCOPE

In this pandemic situation, various health organizations and research centers are trying to collect various samples and their related symptoms from patients suffering from COVID-19. They try to connect with other people to seek information and public opinions about the services and resources available and which are required by them to face this time. If we compare the current situation with past years then we observe that when any problem was faced by a particular region or society at that time, people discuss with each other or with friends and neighbors, but due to advancement of technology, the manner of discussion has changed totally. Now people seek information from the internet and multiple platforms are facilitating social media users according to their required information, which helps them to take the right decision at the right time.

Due to these available resources and facilities, the industrial aspects and business activities have also changed and flourished according to customers' demands and needs. The COVID-19 pandemic has touched every region, part, and corner of the world and spread its impact to almost every possible area. Due to these multiple activities, social platforms are active in these days of the pandemic to update the online social media users and allow them to share their views in terms of text, images, or other possible ways.

Social media users' sentiments and opinions help other users to analyze the current situation with such emerging applications of machine learning in COVID-19 in terms of sentiment analysis (Liu et al., 2007), and event detection (Tumasjan et al., 2010), classification (Asur and Huberman, 2010), and extraction (Yano and Smith, 2010) related to COVID-19 in real-time situations can be analyzed to build a social relation (Groh and Hauffa, 2011) and exchange views and strategies (Zhang and Skiena, 2010). In the following section, a few approaches based on machine learning for COVID-19 text analysis are explained:

- *COVID-19 sentiment analysis based on machine learning approaches (supervised and unsupervised)*: Sentiment analysis research is growing rapidly, and most of the sentiment analysis areas are based on machine learning approaches in which both supervised and unsupervised approaches are included. In general, if we consider COVID-19 sentiment analysis then it is concerned with text classification, which is keyword-centric and based on the key features related to the particular topic. It is similar to topic modeling, where keyword features are established and machine learning approaches are designed to analyze the processed labeled data.

    Therefore, in COVID-19, both supervised and unsupervised approaches can be used to classify the collected COVID-19 text based on their required keyword features. Figure 6.6 illustrates a supervised machine learning approach which includes various phases. In the first phase, various preprocessings, feature selection classifications, and their representations are explored using training documents. Afterward, a machine learning approach is implemented which helps to extract the hidden features based on prediction class. Sometimes it becomes critical to select and analyze features from the collected text as it is concerned with the target area to generate features class. This feature generation helps the machine learning features to be more effective and precise, which help to lessen redundancy and improve performance of the classification task.

Research of COVID-19 text analysis based on their domain and associated features are rigorously dependent on the topic of interest. In the collected text stream of COVID-19, we found that it is a combination of multiple languages where the users are writing in their native languages using slang and local words to address the situation. On the other hand, if we consider the global text, then it is multilingual including multiple languages together. In previous literature of text analysis, classification language-based models are used for analyzing domain-specific features for a particular language, such as in the work by Tan and Zhang (2008), where sentiment analysis is performed on the Chinese language to gain information based on required features.

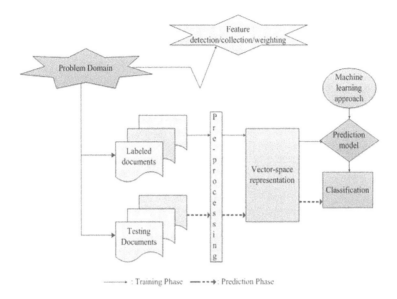

FIGURE 6.6 Depiction of supervised machine learning.

On the other hand, Sharma and Dey (2012) used a corpus of English movies to analyze the sentiment in which feature-based mechanism is used to analyze features. Diverse language feature-based approaches are used by Abbasi et al. (2008) and inter-opinion analysis is addressed by Liang et al. (2015) using matrix factorization. Term Freequency-Inverse document frequency (TF-IDF) is another effective approach which helps to increase the performance and efficiency of classification task (Paltoglou and Thelwall, 2010), and it also helps to retrieve parameters from collected text and calculate the term weight including the whole corpus. In the COVID-19 text classification, sentiment polarity classification is a challenging task as a tweet usually contains hashtags, @ tags, and their relationships. If we analyze the features of COVID-19 using a graphical approach then it can be easy to analyze, because from the root node we can classify and track the symptoms, history, and the other features associated with it. It also helps to predict the social influence and its impact on different domains. Therefore, in such a scenario, expressing sentiments and opinions can help to analyze the features in detail, including their aspect ratio.

If we consider probability-based approaches to analyze sentiments and features then unsupervised machine learning approaches are used, which usually deal with unlabeled text. In this unsupervised approach, it is necessary to capture the domain knowledge of the classification process and the prior knowledge to categorize the sentiments into various sentimental classes (Figure 6.7). To express an opinion and to classify opinion, features-based approach based on part of speech (POS) is applied (Turney, 2002). So we can say that both supervised and unsupervised machine learning approaches can be used but their selection and usage are directly linked with the problem statement and application.

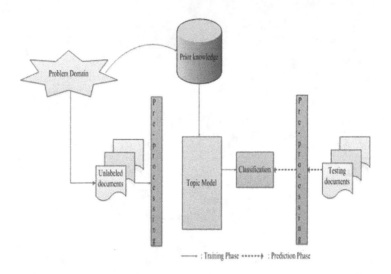

FIGURE 6.7 Depiction of unsupervised machine learning.

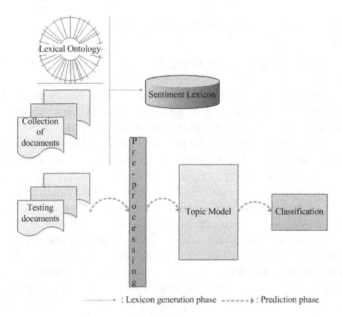

FIGURE 6.8 Depiction of lexicon-based sentiment classification.

- *COVID-19 text analysis using lexicon-based approaches*: Lexicon-based approaches are also used in text analysis. Figure 6.8 illustrates the various resources of lexicons which can be used to understand and classify the features of COVID-19 text. On the other hand, if sentimental keywords available in the collected text are to be classified based on the lexicon features, including their polarity, then this approach is the best. It is observed that the polarity score of sentiments related to COVID-19 is a major problem and we know that it's related to a certain keyword, entity, or feature including their state of conflict. Polarity can be measured on the basis of two phases, like domain dependent and domain independent. On the basis of sentiment lexicons, context prior probability and the contextual polarity are computed to guess the sentiment class. Scores of sentiment classes are computed and feature keywords can be compared with WordNet (Zaśko-Zielińska et al., 2015), SentiWordNet (Esuli and Sebastiani, 2005), or lexicon-based corpus from which the exact meaning of featured keywords can be analyzed. Corpus-based and dictionary-based approaches can be useful to analyze lexicons because lexicons can be created automatically, whereas in dictionary-based methods, word sense can be analyzed by predicting the synonyms and antonyms of the feature keywords (Esuli and Sebastiani, 2005; Hu and Liu, 2004) for COVID-19.

- *Emotion-based COVID-19 text analysis*: Emotion detection, classification, and analysis in COVID-19 are related to the sentimental emotion detection problem domain, which analyzes the text of emotions bearing sentiments related to COVID-19. But in emotion detection, contextual information plays an important role which is to be considered during the analysis of various sentimental keywords. Let us consider the example, "Usually, I don't like the idea to wear mask." This is categorized as positive because it contains positive sentiment words ("like," "idea"), but the negation ("don't") words renders the text negative. Additional machine learning-based classifiers are also used to classify text, and Al-Mannai et al. (2014) designed a hybrid lexicon feature-based framework for sentiment classification. Machine learning approaches are widely used and adopted due to their advanced tendency to analyze domain-based features.

In this pandemic, news channels are also contributing effectively by providing real-time information to viewers, and machine learning can be used to analyze the impact of news in the online news market. From literature, we observed that various authors used the news market to understand the current scenario and the future of the market. Usually they use this social news market prediction for the financial section, as explained by Devitt and Ahmad (2007). So if we consider this news environment for COVID-19 prediction, then it can be successful in analyzing and predicting its future, but one point to be noted is that that the news source should be reliable and its dimensions of correlation with other related indicators should be analyzed. Because of the COVID-19 pandemic, every sector is affected, as many people have been fired by their employers, all countries are facing major economic recession, and every sector of income is in danger.

This COVID-19 pandemic has brought the demise of many sources of income from small-scale industries, which is very difficult to analyze. Therefore, it is necessary to calculate the financial market to predict financial loss by considering multiple parameters. Therefore, in the next section, we will explain the various sectors and areas which are in danger due to this COVID-19 pandemic.

6.8 PREDICTION AND ANALYSIS OF THE IMPACT OF COVID-19 ON DIFFERENT PARAMETERS

COVID-19 appears to have originated in Wuhan City in the Hubei Province of China in December 2019. But now, COVID-19 has been diffused to the whole world in a very short span of time. Every country is struggling to control its impact as no vaccine has been invented yet. But day by day the virus threat is increasing in the worst way possible. Due to this pandemic, almost every country in the world implemented lockdown. Many countries mandated their people to endure lockdowns by stopping all spiritual, national, and socioeconomic events. Many people all over the world have lost their lives and hundreds of thousands of people have been infected, and the trend is still continuing. COVID-19 has had many impacts and has badly affected many different sectors in all countries, of which a few are addressed below:

- *Chaotic sector*: This pandemic has affected workers in a different way as the daily wager employees or those employees who were working in Micro, Small, and Medium Enterprises (MSMEs) have lost everything due to COVID-19. Everywhere there is a sharp decline in job opportunities and a rapid increase in unemployment rates. For micro- and small-scale operations, this disease left no other source of income. This situation has led to starvation and has turned into a very alarming situation for big, developing countries like India, where the situation is becoming critical day by day. People may survive COVID-19, but will definitely die from starvation.
- *Agriculture and food processing*: Agriculture is considered the backbone and a major source of income in most countries. This disease has become a problematic situation for the food processing and agriculture businesses as they are not able to transfer their goods from one state to another and from one country to another country. Farmers and food processing units are incurring huge losses and are forced to throw away their crops. They don't have any other source of income. On social media, we have seen that people are sharing their painful and critical situation. Due to this miserable situation, many people die by suicide and many are suffering from depression as there is no source of income and no hope to thrive again in future.
- *E-commerce*: This critical situation of COVID-19 has put them in a grave situation. The countrywide lockdown will upset the processes of the e-commerce business extremely, particularly at a time when there is a massive need for home delivery of goods. Therefore, this loss of e-commerce industry can be calculated by applying a wide range of machine learning prediction algorithms.
- *Education*: Due to the COVID-19 pandemic, many schools and instructive organizations have been closed down to avoid the transmission of illness among schoolgoers. Though we are protecting their lives, this will also destructively influence their education and academic growth. This year, parents are not willing to send their children to school/college/university. Most educational institutes have shifted from traditional teaching to online teaching. COVID-19 will leave a big impact on the educational areas

and it will later affect admission of students into private and government universities.
- *Tourism and hospitality sector*: Tourism is the backbone of most of the countries. Due to the restrictions on traveling, the revenue of the tourism sector has gone down. Yet most countries have not started their international flights and many tourists' visits have been canceled. Lots of meetings, educational conferences, business meetings, and sports activities, both national and international, including the Olympics, have been canceled or postponed, leading to huge financial losses for different countries.
- *Healthcare industry*: COVID-19 has exposed the liabilities of healthcare structures. We know that availability of healthcare is a fundamental right, but the ubiquitous terror of COVID-19 has in turn affected people's primary healthcare requirements. For even a normal checkup, it has become very difficult to approach a doctor. The situation is more critical for pregnant women, who need to visit obstetricians for prenatal checkups. Many people who are suffering from major health problems are ignored or are left untreated by doctors, and many have lost their lives in this way. So, it is a very saddening situation for every country, where in the 21st century, we say that we have good health facilities.
- *Defense and security*: COVID-19 has jammed the stream chains and production/manufacturing amenities of defense businesses. Due to this, every kind of business is going down and the situation is not in favor of business development. COVID-19 has taught a lesson to the business industry that they need to discover the diverse phases of risk planning. They need to shift themselves on the way to scientific and technical platforms to manage and to handle every critical situation in the future.

Therefore, in the end we can say that almost all sectors have been affected by COVID-19. This situation is becoming more critical every day as the unemployment rate is increasing and resources are diminishing.

6.9 CONCLUSION

This chapter is an endeavor to elucidate text analytics based on machine learning-based text mining in social media with the current, modernized text mining method for the COVID-19 pandemic. As social media has progressed in an exceptional way, it has led to numerous motivating and exciting research directions, particularly in the area of COVID-19. Looking at the interestingness of the text mining area in COVID-19, it is full of information. The foremost aim of this chapter is to expand and conceptualize the areas of text mining in social media which are available and can be accessible to an astonishing extent. Social media text is full of noise, abbreviations, special symbols, emoticons, out-of-vocabulary words, and out of vocabulary words are part of social media text. Therefore, rich patterns of text stream or corpus can be exploited to generate relevant and required information. A number of techniques are present for the COVID-19 text classification based on machine learning, which can't be ignored and can help to measure the impact on text pre-processing and COVID-19 sentiment analysis. In this chapter, we also explained the various tasks along with different application areas. It also discusses the various methodologies available for text pre-processing as well as for sentiment analysis in the text corpus and text stream for COVID-19. As a discovery, these techniques can be used for a number of tasks, and text mining can be applied for the prolific research domain of the COVID-19 disease. It is also observed that the multilingual model which is capable of handling multilingual eccentricities in text stream and text corpus for COVID-19 research can facilitate in a better way the classification and prediction task.

REFERENCES

Abbasi, Ahmed, Hsinchun Chen, and Arab Salem. "Sentiment analysis in multiple languages: feature selection for opinion classification in web forums." *ACM Transactions on Information Systems (TOIS)* 26, no. 3 (2008): 1–34.

Aggarwal, Charu C. "A survey of stream clustering algorithms." In *Proceeding of Data Clustering-Book Serie*, Edition 1, pp. 231–258. Taylor and Francis, 2014

Ahmad, Farooq, and Grzegorz Kondrak. "Learning a spelling error model from search query logs." In *Proceedings of Human Language Technology Conference and Conference on Empirical Methods in Natural Language Processing*, pp. 955–962. 2005.

Aichner, Thomas, and Frank Jacob. "Measuring the degree of corporate social media use." *International Journal of Market Research* 57, no. 2 (2015): 257–276.

Aiello, Luca Maria, Georgios Petkos, Carlos Martin, David Corney, Symeon Papadopoulos, Ryan Skraba, AyseGöker, IoannisKompatsiaris, and Alejandro Jaimes. "Sensing trending topics in Twitter." *IEEE Transactions on Multimedia* 15, no. 6 (2013): 1268–1282.

Al-Mannai, Kamla, Hanan Alshikhabobakr, Sabih Bin Wasi, Rukhsar Neyaz, Houda Bouamor, and Behrang Mohit. "Cmuq-hybrid: sentiment classification by feature engineering and parameter tuning." In *Proceedings of the 8th International Workshop on Semantic Evaluation*, pp. 181–185. 2014.

Alves, Sofia, João Costa, and Jorge Bernardino. "Information extraction applications for clinical trials: a survey." In *2019 14th Iberian Conference on Information Systems and Technologies (CISTI)*, pp. 1–6. IEEE, 2019.

Asur, S., and B.A. Huberman. "Predicting the future with social media." In *2010 IEEE/WIC/ACM International Conference on Web Intelligence and Intelligent Agent Technology*. 2010.

Aw, AiTi, Min Zhang, Juan Xiao, and Jian Su. "A phrase-based statistical model for SMS text normalization." In *Proceedings of the COLING/ACL 2006 Main Conference Poster Sessions*, pp. 33–40. 2006.

Baldwin, Timothy, Paul Cook, Marco Lui, Andrew MacKinlay, and Li Wang. "How noisy social media text, how different social media sources?." In *Proceedings of the Sixth International Joint Conference on Natural Language Processing*, pp. 356–364. 2013.

Bhadane, Chetashri, Hardi Dalal, and Heenal Doshi. "Sentiment analysis: measuring opinions." *Procedia Computer Science* 45 (2015): 808–814.

Bhuta, Sagar, Avit Doshi, Uehit Doshi, and Meera Narvekar. "A review of techniques for sentiment analysis of Twitter data." In *2014 International Conference on Issues and Challenges in Intelligent Computing Techniques (ICICT)*, pp. 583–591. IEEE, 2014.

Choudhury, Monojit, Rahul Saraf, Vijit Jain, Animesh Mukherjee, Sudeshna Sarkar, and Anupam Basu. "Investigation and modeling of the structure of texting language." *International Journal on Document Analysis and Recognition (IJDAR)* 10, no. 3–4 (2007): 157–174.

Chrupała, Grzegorz. "Normalizing tweets with edit scripts and recurrent neural embeddings." In *Proceedings of the 52nd Annual Meeting of the Association for Computational Linguistics (Volume 2: Short Papers)*, pp. 680–686. 2014.

Cook, Paul, and Suzanne Stevenson. "An unsupervised model for text message normalization." In *Proceedings of the Workshop on Computational Approaches to Linguistic Creativity*, pp. 71–78. 2009.

Croft, W. Bruce, Donald Metzler, and Trevor Strohman. *Search Engines: Information Retrieval in Practice*. vol. 520. Reading: Addison-Wesley, 2010.

Cucerzan, Silviu, and Eric Brill. "Spelling correction as an iterative process that exploits the collective knowledge of web users." In *Proceedings of the 2004 Conference on Empirical Methods in Natural Language Processing*, pp. 293–300. 2004.

Daume III, Hal, and Daniel Marcu. "Domain adaptation for statistical classifiers." *Journal of Artificial Intelligence Research* 26 (2006): 101–126.

Derczynski, Leon, Diana Maynard, Giuseppe Rizzo, Marieke Van Erp, Genevieve Gorrell, Raphaël Troncy, Johann Petrak, and Kalina Bontcheva. "Analysis of named entity recognition and linking for tweets." *Information Processing & Management* 51, no. 2 (2015): 32–49.

Desikan, Prasanna, Colin DeLong, Sandeep Mane, Kalyan Beemanapalli, Kuo-Wei Hsu, Prasad Sriram, Jaideep Srivastava, and Vamsee Venuturumilli. "Web mining for business computing." *Business Computing* 3 (2006): 45–71.

Devitt, Ann, and Khurshid Ahmad. "Sentiment polarity identification in financial news: a cohesion-based approach." In *Proceedings of the 45th Annual Meeting of the Association of Computational Linguistics*, pp. 984–991. 2007.

Emadi, Mehdi, and Maseud Rahgozar. "Twitter sentiment analysis using fuzzy integral classifier fusion." *Journal of Information Science* 46, no. 2 (2020): 226–242.

Esuli, Andrea, and Fabrizio Sebastiani. "Determining the semantic orientation of terms through gloss classification." In *Proceedings of the 14th ACM International Conference on Information and Knowledge Management*, pp. 617–624. 2005.

Foster, Jennifer. "'cba to check the spelling': investigating parser performance on discussion forum posts." In *Human Language Technologies: The 2010 Annual Conference of the North American Chapter of the Association for Computational Linguistics*, pp. 381–384. 2010.

Gelfand, Boris, Marilyn Wulfekuler, and W.F. Punch. "Automated concept extraction from plain text." In *AAAI 1998 Workshop on Text Categorization*, pp. 13–17. 1998.

Ghai, Deepika, Divya Gera, and Neelu Jain. "A new approach to extract text from images based on DWT and K-means clustering." *International Journal of Computational Intelligence Systems* 9, no. 5 (2016): 900–916.

Gimpel, Kevin, Nathan Schneider, Brendan O'Connor, Dipanjan Das, Daniel Mills, Jacob Eisenstein, Michael Heilman, Dani Yogatama, Jeffrey Flanigan, and Noah A. Smith. *Part of speech Tagging for Twitter: Annotation, Features, and Experiments*. Carnegie-Mellon Univ Pittsburgh Pa School of Computer Science, 2010.

Goldhahn, Dirk, Thomas Eckart, and Uwe Quasthoff. "Building large monolingual dictionaries at the leipzig corpora collection: from 100 to 200 languages." In *LREC*, vol. 29, pp. 31–43. 2012.

Groh, Georg, and Jan Hauffa. "Characterizing social relations via nlp-based sentiment analysis." In *Fifth International AAAI Conference on Weblogs and Social Media*, 2011.

Hamdan, Hussam, Patrice Bellot, and Frederic Bechet. "lsislif: feature extraction and label weighting for sentiment analysis in twitter." In *Proceedings of the 9th International Workshop on Semantic Evaluation (SemEval 2015)*, pp. 568–573. 2015.

Han, Bo. *Improving the Utility of Social Media With Natural Language Processing*. PhD diss., 2014.

Han, Bo, Paul Cook, and Timothy Baldwin. "Automatically constructing a normalisation dictionary for microblogs." In *Proceedings of the 2012 Joint Conference on Empirical Methods in Natural Language Processing and Computational Natural Language Learning*, pp. 421–432. 2012.

Hassan, Hany, and Arul Menezes. "Social text normalization using contextual graph random walks." In *Proceedings of the 51st Annual Meeting of the Association for Computational Linguistics (Volume 1: Long Papers)*, pp. 1577–1586. 2013.

Hu, Minqing, and Bing Liu. "Mining and summarizing customer reviews." In *Proceedings of the Tenth ACM SIGKDD International Conference on Knowledge Discovery and Data Mining*, pp. 168–177. 2004.

Kaufmann, Max, and Jugal Kalita. "Syntactic normalization of twitter messages." In *International Conference on Natural Language Processing*, Kharagpur, India, vol. 16. 2010.

Klein, Dan, and Christopher D. Manning. "Accurate unlexicalized parsing." In *Proceedings of the 41st Annual Meeting of the Association for Computational Linguistics*, pp. 423–430. 2003.

Koehn, Philipp, Hieu Hoang, Alexandra Birch, Chris Callison-Burch, Marcello Federico, Nicola Bertoldi, Brooke Cowan et al. "Moses: open source toolkit for statistical machine translation." In *Proceedings of the 45th Annual Meeting of the ACL on Interactive Poster and Demonstration Sessions*, pp. 177–180. Association for Computational Linguistics, 2007.

Kukich, Karen. "Techniques for automatically correcting words in text." *ACM Computing Surveys (CSUR)* 24, no. 4 (1992): 377–439.

Kumar, J. Prasanna, and Paladugu Govindarajulu. "Near-duplicate web page detection: an efficient approach using clustering, sentence feature and fingerprinting." *International Journal of Computational Intelligence Systems* 6, no. 1 (2013): 1–13.

Li, Mu, Muhua Zhu, Yang Zhang, and Ming Zhou. "Exploring distributional similarity based models for query spelling correction." In *Proceedings of the 21st International Conference on Computational Linguistics and 44th Annual Meeting of the Association for Computational Linguistics*, pp. 1025–1032. 2006.

Liang, J., X. Zhou, L. Guo, and S. Bai, "Feature selection for sentiment classification using matrix factorization categories and subject descriptors." In *WWW '15 Companion: Proceedings of the 24th International Conference on World Wide Web*, pp. 63–64. 2015.

Lifna, C.S., and M. Vijayalakshmi. "Identifying concept-drift in twitter streams." *Procedia Computer Science* 45 (2015): 86–94.

Lin, Dekang. "Automatic retrieval and clustering of similar words." In *36th Annual Meeting of the Association for Computational Linguistics and 17th International Conference on Computational Linguistics*, vol. 2, pp. 768–774. 1998.

Lin, Jimmy, Rion Snow, and William Morgan. "Smoothing techniques for adaptive online language models: topic tracking in tweet streams." In *Proceedings of the 17th ACM SIGKDD International Conference on Knowledge Discovery and Data Mining*, pp. 422–429. 2011.

Liu, Bing. "Sentiment analysis and opinion mining." *Synthesis Lectures on Human Language Technologies* 5, no. 1 (2012): 1–167.

Liu, Bing. "Sentiment analysis and subjectivity." *Handbook of Natural Language Processing* 2 (2010): 627–666.

Liu, Xiaohua, Shaodian Zhang, Furu Wei, and Ming Zhou. "Recognizing named entities in tweets." In *Proceedings of the 49th Annual Meeting of the Association for Computational Linguistics: Human Language Technologies*, pp. 359–367. 2011.

Liu, Yang, Xiangji Huang, Aijun An, and Xiaohui Yu. "ARSA: a sentiment-aware model for predicting sales performance using blogs." In *Proceedings of the 30th Annual International ACM SIGIR Conference on Research and Development in Information Retrieval*, pp. 607–614. 2007.

Marcus, Mitchell, Beatrice Santorini, and Mary Ann Marcinkiewicz. *Building a Large Annotated Corpus of English*. The Penn Treebank, 1993.

Martínez, Luis, Manuel J. Barranco, Luis G. Pérez, and Macarena Espinilla. "A knowledge based recommender system with multigranular linguistic information." *International Journal of Computational Intelligence Systems* 1, no. 3 (2008): 225–236.

Owoputi, Olutobi, Brendan O'Connor, Chris Dyer, Kevin Gimpel, Nathan Schneider, and Noah A. Smith. "Improved part-of-speech tagging for online conversational text with word clusters." In *Proceedings of the 2013 Conference of the North American Chapter of the Association for Computational Linguistics: Human Language Technologies*, pp. 380–390. 2013.

Paltoglou, Georgios, and Mike Thelwall. "A study of information retrieval weighting schemes for sentiment analysis." In *Proceedings of the 48th Annual Meeting of the Association for Computational Linguistics*, pp. 1386–1395. 2010.

Pang, Bo, and Lillian Lee. "Opinion mining and sentiment analysis." *Foundations and Trends in Information Retrieval* 2 (2008): 1–2.

Pang, Bo, Lillian Lee, and Shivakumar Vaithyanathan. "Thumbs up? Sentiment classification using machine learning techniques." *arXiv Preprint cs/0205070* (2002).

Peetz, Maria-Hendrike. *Time-aware Online Reputation Analysis*. Universiteit van Amsterdam [Host], 2015.

Pennell, Deana, and Yang Liu. "A character-level machine translation approach for normalization of sms abbreviations." In *Proceedings of 5th International Joint Conference on Natural Language Processing*, pp. 974–982. 2011.

Pomares Quimbaya, Alexandra, Alejandro Sierra Múnera, Rafael Andrés González Rivera, Julián Camilo Daza Rodríguez, Oscar Mauricio Muñoz Velandia, Angel Alberto García Peña, and Cyril Labbé. "Named entity recognition over electronic health records through a combined dictionary-based approach." In *Procedia Computer Science*, pp. 55–61. Elsevier, CENTERIS, 2016.

Popowich, Fred. "Using text mining and natural language processing for health care claims processing." *ACM SIGKDD Explorations Newsletter* 7, no. 1 (2005): 59–66.

Quan, Changqin, and Fuji Ren. "Unsupervised product feature extraction for feature-oriented opinion determination." *Information Sciences* 272 (2014): 16–28.

Ren, Zhaochun. *Monitoring Social Media: Summarization, Classification and Recommendation*. Universiteit van Amsterdam [Host], 2016.

Ren, Zhaochun, Shangsong Liang, Edgar Meij, and Maarten de Rijke. "Personalized time-aware tweets summarization." In *Proceedings of the 36th International ACM SIGIR Conference on Research and Development in Information Retrieval*, pp. 513–522. 2013.

Ritter, Alan, Sam Clark, and Oren Etzioni. "Named entity recognition in tweets: an experimental study." In *Proceedings of the 2011 Conference on Empirical Methods in Natural Language Processing*, pp. 1524–1534. 2011.

Saleiro, Pedro, Eduarda Mendes Rodrigues, Carlos Soares, and Eugénio Oliveira. "Feup at semeval-2017 task 5: predicting sentiment polarity and intensity with financial word embeddings." *arXiv Preprint arXiv:1704.05091* (2017).

Santana, Adamo, Souta Inoue, Kenya Murakami, Tatsuya Iizaka, and Tetsuro Matsui. "Clustering-based data reduction approach to speed up SVM in classification and regression tasks." In *International Conference on Industrial, Engineering and Other Applications of Applied Intelligent Systems*, pp. 478–488. Springer, Cham, 2020.

Sarker, Abeed. "A customizable pipeline for social media text normalization." *Social Network Analysis and Mining* 7, no. 1 (2017): 45.

Schlippe, Tim, Chenfei Zhu, Jan Gebhardt, and Tanja Schultz. "Text normalization based on statistical machine translation and internet user support." In *Eleventh Annual Conference of the International Speech Communication Association*. 2010.

Sharma, Anuj, and Shubhamoy Dey. "Performance investigation of feature selection methods and sentiment lexicons for sentiment analysis." *IJCA Special Issue on Advanced Computing and Communication Technologies for HPC Applications* 3 (2012): 15–20.

Singh, Tajinder, and Madhu Kumari. "Role of text pre-processing in twitter sentiment analysis." *Procedia Computer Science* 89, no. Supplement C (2016): 549–554.

Sproat, Richard, Alan W. Black, Stanley Chen, Shankar Kumar, Mari Ostendorf, and Christopher Richards. "Normalization of non-standard words." *Computer Speech & Language* 15, no. 3 (2001): 287–333.

Tan, Songbo, and Jin Zhang. "An empirical study of sentiment analysis for Chinese documents." *Expert Systems With Applications* 34, no. 4 (2008): 2622–2629.

Toutanova, Kristina, Dan Klein, Christopher D. Manning, and Yoram Singer. "Feature-rich part-of-speech tagging with a cyclic dependency network." In *Proceedings of the 2003 Conference of the North American Chapter of the Association for Computational Linguistics on Human Language Technology-Volume 1*, pp. 173–180. Association for Computational Linguistics, 2003.

Tumasjan, Andranik, Timm O. Sprenger, Philipp G. Sandner, and Isabell M. Welpe. "Predicting elections with twitter: what 140 characters reveal about political sentiment." In *Fourth International AAAI Conference on Weblogs and Social Media*. 2010.

Turney, Peter D. "Thumbs up or thumbs down? Semantic orientation applied to unsupervised classification of reviews." *arXiv Preprint cs/0212032* (2002).

Vavliakis, Konstantinos N., Andreas L. Symeonidis, and Pericles A. Mitkas. "Event identification in web social media through named entity recognition and topic modeling." *Data & Knowledge Engineering* 88 (2013): 1–24.

Wang, Pidong, and Hwee Tou Ng. "A beam-search decoder for normalization of social media text with application to machine translation." In *Proceedings of the 2013 Conference of the North American Chapter of the Association for Computational Linguistics: Human Language Technologies*, pp. 471–481. 2013.

Wei, Bin, and Christopher Pal. "Cross lingual adaptation: an experiment on sentiment classifications." In *Proceedings of the ACL 2010 Conference Short Papers*, pp. 258–262. 2010.

Whitelaw, Casey, Ben Hutchinson, Grace Chung, and Ged Ellis. "Using the web for language independent spellchecking and autocorrection." In *Proceedings of the 2009 Conference on Empirical Methods in Natural Language Processing*, pp. 890–899. 2009.

Wilcox-O'Hearn, L. Amber. "Detection is the central problem in real-word spelling correction." *arXiv Preprint arXiv:1408.3153* (2014).

Yang, Yi, and Jacob Eisenstein. "A log-linear model for unsupervised text normalization." In *Proceedings of the 2013 Conference on Empirical Methods in Natural Language Processing*, pp. 61–72. 2013.

Yano, Tae, and Noah A. Smith. "What's worthy of comment? content and comment volume in political blogs." In *Fourth International AAAI Conference on Weblogs and Social Media*. 2010.

Zaśko-Zielińska, Monika, Maciej Piasecki, and Stan Szpakowicz. "A large wordnet-based sentiment lexicon for polish." In *Proceedings of the International Conference Recent Advances in Natural Language Processing*, pp. 721–730. 2015.

Zhang, Congle, Tyler Baldwin, Howard Ho, Benny Kimelfeld, and Yunyao Li. "Adaptive parser-centric text normalization." In *Proceedings of the 51st Annual Meeting of the Association for Computational Linguistics (Volume 1: Long Papers)*, pp. 1159–1168. 2013.

Zhang, Wenbin, and Steven Skiena. "Trading strategies to exploit blog and news sentiment." In *Icwsm*, 2010.

Zhao, Qiankun, and Prasenjit Mitra. "Event detection and visualization for social text streams." In *ICWSM*, 2007.

Zhou, Xiangmin, and Lei Chen. "Event detection over twitter social media streams." *VLDB Journal* 23, no. 3 (2014): 381–400.

7 Containing the Spread of COVID-19 with IoT
A Visual Tracing Approach

Pallav Kumar Deb, Sudip Misra,
Anandarup Mukherjee, and Aritra Bandyopadhyay

CONTENTS

7.1	Introduction	128
	7.1.1 Motivation	130
7.2	Overview of Techniques for Combating COVID-19	130
	7.2.1 COVID-19 and IoT	131
	7.2.2 Computer Vision	131
	7.2.3 Human Activity Recognition	131
	7.2.4 Fog/Edge Computing	132
	7.2.5 Synthesis	132
7.3	System Model	132
	7.3.1 IoT-Based Network Architecture for COVI-SCANNER	133
	7.3.2 Information Flow in COVI-SCANNER	133
	7.3.3 Proposed Solution	133
	7.3.3.1 The Contamination Module	134
	7.3.3.2 The Sanitization Module	135
7.4	Performance Evaluation	135
	7.4.1 Experimental Setup	136
	7.4.2 Results	136
	7.4.2.1 Output from Contamination Phase	136
	7.4.2.2 Output from Sanitization Phase	137
	7.4.2.3 Detection and Removal of Fomite Spaces	137
	7.4.2.4 Accuracy of the COVI-SCANNER Phases	138
	7.4.2.5 Delays in Executing COVI-SCANNER and Its Phases	139
	7.4.2.6 Upload and Download Rates	140
7.5	Conclusion	142
References		142

7.1 INTRODUCTION

The outbreak of the COVID-19 virus has spread at an alarming rate all around the world. The intangible nature of the virus and its transmission mode make containment challenging. Moreover, the droplets from an infected individual remain remnant in their hands in addition to transmission in the form of aerosols, which leads to contamination of public objects and spaces on interaction (touching and passage). The COVID-19 virus survives for almost 3 days on surfaces made of steel and plastic, and over 24 hours on cardboards (Chamola et al., 2020). This gives rise to fomite spaces, and such intractable contamination rapidly facilitates secondary spreading, especially in closed public environments. As a healthy person interacts with these fomite spaces, the virus enters into the body, binds itself to cellular receptors, and starts multiplying, sometimes leading to fatality. Such a high risk of secondary transmission mandates the need for rigorous contact tracing (Gan et al., 2020). In such scenarios, Internet of Things (IoT)-based solutions have the potential to combat the COVID-19 virus by tracking and containing fomite spaces. The use of smart solutions based on image and video processing for tracking fomite spaces is beneficial in restricting secondary spreading in closed environments. Such image processing-based solutions require high configuration devices for seamless execution. However, adopting legacy and affordable IoT infrastructures to deploy such monitoring technologies is beneficial.

In this chapter, we propose COVI-SCANNER, an augmented reality (AR)-based fomite space monitor in closed environments on a fog-enabled IoT platform. As shown in Figure 7.1, the COVI-SCANNER scheme works in two phases: (1) *contamination*

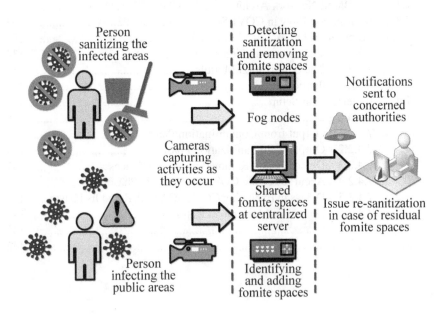

FIGURE 7.1 Overview of the proposed COVI-SCANNER scheme.

and (2) *sanitization*. In the contamination phase, COVI-SCANNER records the videos from the cameras for tracking the movement of infected individuals and detects the spaces/objects that the person touches. On detecting any such contact, the COVI-SCANNER highlights them (virtually) as potentially contaminated/fomite places. In the sanitization phase, the COVI-SCANNER tries to look for cleaning materials such as brooms, wipes, and mops. Upon detection of these objects, the COVI-SCANNER tries to detect cleaning activity by the person interacting with the sanitization objects. If the person starts cleaning and the object passes through the fomite places identified during the contamination phase, it removes the demarcations. If the cleaning activity is complete and all of the demarcations are not removed at the end of the sanitization phase, COVI-SCANNER raises an alarm and notifies the concerned authorities. This method allows us to ensure the existence of minimal fomite places in public areas and reduce the secondary spread. Toward this, we use the readily available machine learning models: (1) MobileNet-SSD (Younis et al., 2020), for detecting the person and objects in the contamination phase, and (2) 3D-ResNet (Hara et al., 2018), for identifying the cleaning activity in the sanitization phase. It may be noted that we assume that the concerned authorities have information on the infected individuals and recognize them using available face recognition techniques (Khan et al., 2019). We further propose a fog-based IoT architecture to deploy the proposed COVI-SCANNER scheme on pre-installed infrastructures. We account for the computational complexity in the AR-based operations and propose assigning different fog nodes for each contamination and sanitization phase. Since the two phases share the information on the fomite spaces, we propose storing the points in a shared database with provisions for real-time updates on highlighting and removal from the two phases. In this chapter, we propose COVI-SCANNER, an AR-based fomite space monitor in closed environments to minimize the secondary spreading of the COVID-19 virus. For the ease of deployment on legacy infrastructures, we propose a fog-enabled IoT architecture comprising of resource-constrained devices. The major highlights of this chapter are as follows:

- *COVI-SCANNER*: We propose an AR-based method for tracking and monitoring fomite spaces to reduce the secondary spread of the COVID-19 virus.
- *Modular operations*: As image processing techniques involve complex computations, we divide COVI-SCANNER into two phases: *contamination* and *sanitization*. Such modules help in reducing the load from the devices.
- *IoT-Based architecture*: To facilitate easy deployment on resource-constrained devices, we propose an IoT-based architecture that assigns dedicated fog nodes for performing each phase and sharing their data from a common database server in real time.
- *Evaluation*: To show the feasibility of the proposed COVI-SCANNER scheme, we implement and deploy in lab-scale and present the observed results.

Example scenario: Consider a closed environment such as hospitals, industries, and any other environment with cameras installed for monitoring. In such environments, fog nodes assigned for the contamination phase in COVI-SCANNER identifies the infected individuals. It then creates a bounding box around the person and tracks the person, particularly the arms and their interactions with nearby objects. In the case of detecting such interactions, COVI-SCANNER marks them as fomite spaces on the screen. On the other hand, fog nodes assigned for the sanitization phase in COVI-SCANNER first identify the necessary cleaning objects such as wipes, mops, and brooms to then detect the cleaning activity. As the individual cleans the regions, COVI-SCANNER tracks the cleaning objects and removes the highlighted portions. The COVI-SCANNER notifies the concerned authorities in case all the fomite spaces are not sanitized. The fog nodes operating under each phase share their data through a central database server for simultaneously updating the data in real time.

7.1.1 Motivation

The recent outbreak of the COVID-19 virus has spread rapidly all around the world. The intangible nature of the virus and its mode of spread of the droplets in the form of aerosols and contact have increased the challenges of containing it. Current solutions depend on contact tracing and self-assessment tests by individuals (Menni et al., 2020), which is not reliable as they may enter false information to avoid isolation and containment. The remnants of the virus in an infected individual, particularly in the hands, cause contamination of objects and spaces with each touch, giving rise to fomite spaces. Healthy individuals acquire this virus by interacting with the same set of fomite spaces. The virus stays alive on cardboard and plastic surfaces for a duration of 24 hours to 3 days, respectively. Such a lifespan increases the risk of spreading to a greater number of individuals each day, which mandates the need for efficient sanitization. The intractable transmission mode of the virus necessitates IoT-based solutions for monitoring and tracking fomite spaces in closed public areas to reduce secondary spread. Moreover, we propose an AR-based method (COVI-SCANNER) to combat the secondary spread of the virus. However, image processing techniques involve complex operations for execution. Facilitating such operations on legacy infrastructures requires task distribution techniques for seamless deployment. Such issues act as our motivation for designing COVI-SCANNER and its IoT-based architecture. We envision such solutions to help in restricting the secondary spread of the COVID-19 virus in closed public environments.

The rest of the chapter is organized as follows. We present some of the existing works in literature and techniques for combating the COVID-19 virus in Section 7.2. In Section 7.3, we present the system model. We discuss our observations in Section 7.4 and finally conclude in Section 7.5.

7.2 OVERVIEW OF TECHNIQUES FOR COMBATING COVID-19

In this section, we present some of the current literature on the COVID-19 virus and IoT solutions toward combatting it. We then present some of the existing works on computer vision, human activity recognition, and fog/edge computing solutions.

7.2.1 COVID-19 AND IoT

The COVID-19 virus needs a week to show initial symptoms of infection and some do not show any symptoms at all (asymptotic cases). Benreguia et al. (2020) identified the issue and the significant difference in the documented and the actual count of positively infected individuals due to it. They proposed an IoT-based solution to track the known individuals and locations they have been. As healthy individuals visit the same location, the proposed solution warns them of potential infection. Wang et al. (2020) exploited the Social Internet of Things (SIoT) for identifying social relationships and potentially infected individuals as they come in contact with positively infected patients. They used a graph theoretic approach coupled with reinforcement learning to realize the proposed patient identification method. Researchers are also exploring machine learning (ML) methods for combatting the pandemic. Waheed et al. (2020) focused on the lack of data due to the recent outbreak and the major drawback of the ML methods. They developed synthetic X-ray images by considering the effects of the COVID-19 virus and trained an Auxiliary Classifier Generative Adversarial Network (ACGAN) for identifying infected individuals with high precision. Hussain et al. (2020) have provided a comprehensive description of the ML methods and their role in combatting the COVID-19 virus.

7.2.2 COMPUTER VISION

In this section, we highlight some of the works in literature focusing on the development of computer vision techniques and their applications. Morales et al. (2019) used a combination of VGG-16 network and convolutional Long Short Term Memory (LSTM) layers to detect violent robberies through CCTV camera footage. Wong et al. (2020) proposed a novel approach to re-identification of a person on campus using multiple CCTV cameras by assigning higher weights to combinations of parts that helped in re-identification by evaluating the relative performances of each of these combinations. Khandelwal et al. (2020) used face detection and person detection algorithms to raise alarms if people were detected not wearing masks or not following social distancing rules and also implemented the same in manufacturing plants having multiple CCTV cameras.

7.2.3 HUMAN ACTIVITY RECOGNITION

In this chapter, we use human activity recognition to distinguish between a normal person and a person who is cleaning infected areas. This field has seen a lot of research work in recent years mainly due to the increasing popularity of deep learning. Xu et al. (2019) proposed a deep learning model that draws inspiration from the inception network and Gated Recurrent Unit (GRU) to predict human activity by detecting inputs in the form of waveform data from multiple sensors attached to the body. Gnouma et al. (2019) proposed to use a dynamic frame skipping method and used Gaussian Mixture Model for foreground detection, both of which reduced the time taken for silhouette extraction which is required for human activity recognition. Noori et al. (2019) trained a Recurrent Neural Network consisting of Long

Short Term Memory cells on the OpenPose dataset to predict activity performed by a human from different camera angles.

7.2.4 Fog/Edge Computing

Edge computing helps in keeping computational and storage units closer to the devices for reducing bandwidth usage and response time. Abdellatif et al. (2019) discussed the challenges of using edge computing concepts in healthcare systems and discussed in-depth how wearable sensors and medical devices on the edge of the network could be used to monitor the health conditions of patients while ensuring user privacy is maintained as well. Cao et al. (2015) proposed a real-time fall detection system for stroke patients by dividing the computation of analytics between smartphones with accelerometers having lower computational speed and edge nodes with higher computational speed to reduce and rectify false detections. Barthelemy et al. (Barthelemy et al., 2019) proposed an edge computing architecture using a live feed from CCTV cameras across a smart city by using popular lightweight algorithms like YOLO V3 (Redmon and Farhadi 2018) and Simple Online and Realtime Tracking (SORT) algorithm to perform object detection and object tracking, respectively. Deb et al. (2020) proposed a digital stethoscope SkopEdge for counting the number of heartbeats by exploiting the features of edge computing. They proposed sending the recorded audio in the most suitable format based on the network state and the device conditions. IoT devices such as SkopEdge have the potential of addressing the problems in the identification of the COVID-19 virus.

7.2.5 Synthesis

From the discussion in this section, it is evident that IoT solutions play a major role in helping combat the rapidly spreading COVID-19 virus. These solutions help in tracking the positively infected individuals and identifying potentially asymptotic carriers based on the locations that they visit. Additionally, ML methods also help in diagnosing the patients and categorizing them as infected or healthy individuals. We also notice that the popularly available pre-trained ML models based on image processing help in maintaining social distancing and identifying violators. Since ML techniques involve computationally complex operations, fog/edge computing methods help in deploying the trained models on resource-constrained devices. Although we notice a myriad of applications, we observe a lacuna in reducing the secondary spread of the virus from fomite spaces. Toward this, we propose the COVI-SCANNER scheme using an IoT architecture to first identify and highlight the fomite spaces (contamination). Then, we propose identifying cleaning objects and the corresponding activity to trace and remove the highlighted fomite spaces (sanitization).

7.3 SYSTEM MODEL

In this section, we present the IoT network architecture adopted for the proposed COVI-SCANNER scheme along with the information flow. Additionally, we also explain the solution technique for realizing COVI-SCANNER.

7.3.1 IoT-Based Network Architecture for COVI-SCANNER

We consider a set of preinstalled surveillance cameras $C = \{c_1, c_2, \ldots, c_p\}$ responsible for monitoring the closed environments, as shown in Figure 7.1. The images and videos from these cameras pass through fog nodes before reaching the centralized storage server. In this chapter, we consider the routers, switches, and other devices present at the edge of the network as fog nodes. Considering a set of fog nodes $F = \{f_1, f_2, \ldots, f_q\}$, we split them into two subsets for performing each of the two phases in COVI-SCANNER. In other words, $F_c \subset F$ fog nodes for contamination and $F_s \subset F$ fog nodes for sanitization. It may be noted that the fog nodes in F_c and F_s may not be mutually explicit. The fog nodes with high configurations may execute both the modules for contamination and sanitization, respectively. It may be noted that the selection of the optimal set of fog nodes is beyond the scope of this chapter and we rely on available works on resource allocation (Tran and Pompili 2018). Fog nodes executing the contamination routine track the infected individuals and highlight the fomite spaces on the screen. It then sends the details of each of the fomite spaces for storage and access by the fog nodes assigned for the sanitization phase. As the cleaning activity proceeds and the fog nodes detect the sanitization of the fomite spaces, it removes the corresponding information from the server. The data in the server is simultaneously accessible by both sets of fog nodes so that the concerned personnel may view the results on a screen in real time. In case the cleaning activity is over and fomite spaces continue to persist in the region, the server may issue an alarm for notifying the concerned authorities.

7.3.2 Information Flow in COVI-SCANNER

Figure 7.2 depicts the information flow among the devices mentioned in Section 7.3.1. The F_c fog nodes determine the fomite spaces and their coordinates in the contamination phase (step 1). They send the coordinates of the infected regions along with the details of the bounding boxes to the centralized servers (step 2). These coordinates are accessible by the sanitization phase to keep track of the cleaning of the fomite spaces (step 3) and for performing the execution routine (step 4). The fog nodes remove the demarcated fomite spaces as the cleaning activity proceeds (step 5). In the case of residual fomite spaces after sanitization, the server notifies the concerned authorities (step 6). It may be noted that there is no need for manually removing the demarcated fomite spaces as the sanitization phase will automatically remove the same on detecting each sequence of cleaning activities (recurrent step 5).

7.3.3 Proposed Solution

In this section, we describe the proposed solution technique and the adopted pretrained ML models for realizing COVI-SCANNER. Conforming to the network architecture in Section 7.3.1 and the information flow in Section 7.3.2, we present each of the phases in COVI-SCANNER separately in the subsequent sections.

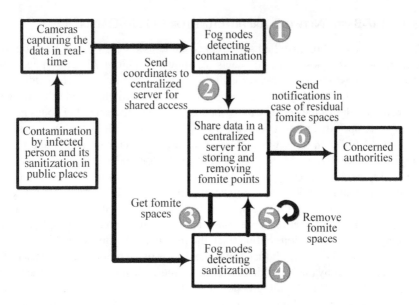

FIGURE 7.2 Information flow in the proposed COVI-SCANNER scheme.

7.3.3.1 The Contamination Module

We execute this module on the F_c set of fog nodes. We use the MobileNet-SSD network available in the deep learning module of OpenCV for object detection. The MobileNet-SSD network detects and returns coordinates of the bounding boxes for 20 classes of objects. Among these 20 classes, we only use the class labeled as *person* for our application and ignore the other classes. As mentioned earlier, we assume that the infected individuals are known beforehand and facial recognition techniques may be used for detecting them on the screen. We use the person class as proof of concept for the proposed COVI-SCANNER scheme. The MobileNet class of convolutional neural networks uses depth-wise separable convolutions followed by 1×1 point-wise convolutions, which allows these models to be much more computationally faster (Howard et al., 2017). This model allows us to detect and draw bounding boxes around all person objects detected in each frame. This module is entirely run on the contamination devices.

Each person detected from the developed model p_i has its own bounding box coordinates, represented by the upper left point (x_1^i, y_1^i) and the lower right point (x_2^i, y_2^i). For demarcating the places where each person p_i has visited and interacted with objects, we calculate a fomite point $fp_i \in \mathbb{R}^2$ for each person object. We calculate the fp_i in such a manner to make sure that the fomite point is nearer to the points in the frame which helps in tracking the cleaning objects during the activity. Mathematically, we calculate fp_i as:

$$fp_i = \left(\frac{x_2^i + x_1^i}{2}, \frac{9y_2^i + y_1^i}{10} \right) \quad (1.1)$$

This fomite point fp_i is used for tracking person objects from one frame to the next. Let the set of all person objects detected in a particular frame j be P^j while the set of all person objects detected in the next frame be P^{j+1}. Let each person object in frame j be $p_i^j \in P^j$ and each person in the next frame $j+1$ be $p_i^{j+1} \in P^{j+1}$. Let the fomite point of each person object p_i^j be fp_i^j, such that $fp_i^j \in F_i^j$. Initially, we calculate the Euclidean distances between all pairs of fomite points in P^j and P^{j+1}. We define each of the distances such as between fp_i^j and fp_i^{j+1} as $dist((fp, fp_i^{j+1}))$. To track each person on the screen effectively, we rely on the following set of rules:

- *Already detected person*: For any fixed p_i^j, we consider the person to be already detected in p_i^{j+1} if $dist((fp_i^j, fp_i^{j+1}))$ is minimum $\forall\ fp_i^{j+1} \in F_i^{j+1}$. We use this to rediscover already detected people from the previous frame j in the current frame $j+1$.
- *New person detection*: We demarcate an object as a new person and add identities on the screen to all objects in P^{j+1} that were not mapped as a person in P^j in the previous step. This is used when a new person who was not present in the frame j has just entered in the next frame $j+1$.
- *Terminate tracking*: If p^{j+1} remains unassigned to any person object in P^j for 10 consecutive frames, then we remove this person p_i from the screen and store the list of all its corresponding fomite points in the centralized server for future reference. This is used to stop trying to find the closest fp_i^{j+1} to a particular fp_i^j, belonging to a person p_i^j who has already exited the area of the frame.

7.3.3.2 The Sanitization Module

We execute this module entirely on the set of sanitization fog node (F_s) devices. We pass the bounding box of each detected person as input to a separate deep learning model to detect whether the person is performing the cleaning activity or not. If the person is detected to have been performing the activity of cleaning then its focus point is calculated. Let this focus point of cleaner person object be f_C. Let $M \in \mathbb{R}^2$ be the set of the history of all fomite points of all persons detected so far. Let m_i be each fomite point in M. To remove the fomite points of people in areas where cleaning activity has occurred, we remove all m_i from M, such that $dist(f_C, m_i) < D_{max}$.

In our implementation of COVI-SCANNER, we consider $D_{max} = 50$ pixels. We use the pre-trained 3D-ResNet deep learning model which is trained on the Kinetics Human Action Video Dataset (Kay et al., 2017) for identifying up to 400 different types of human activities to detect the cleaning activity. Out of all the 400 activity classes, we use only two classes of *cleaning floor* and *mopping floor* and categorize them as cleaning. We do so by taking the previous N frames as input and predicting the activity in the $(N+1)$ th frame. In this chapter, we use $N=16$ and $N=10$ while performing our experiments.

7.4 PERFORMANCE EVALUATION

In this section, we present our lab-scale experimental setup and present our observations on deploying COVI-SCANNER on the proposed IoT architecture.

7.4.1 Experimental Setup

We use two arbitrary systems with i3 and i5 processors and assign them for executing the contamination and sanitization modules. We install the MobileNet-SSD and 3D-ResNet ML models on the concerned devices and use the videos from the cameras as inputs in each case. We use Python 3.7 platform for realizing the proposed COVI-SCANNER scheme and to present its performance.

7.4.2 Results

In this section, we present and discuss the performance of the proposed COVI-SCANNER scheme on deployment. Toward this, we first present its output in identifying and removing the fomite spaces in each phase, followed by its corresponding accuracy. We then demonstrate the delay in executing each phase together with the upload rate for the contamination and sanitization phases along with the download rate at the centralized server.

7.4.2.1 Output from Contamination Phase

We capture one of the instances from the contamination phase. As mentioned earlier, we assume that the identity of the infected individuals is known beforehand and they may be identified using available face recognition techniques. In this chapter, we consider any random person to be infected with the COVID-19 virus for proof of concept. We observe in Figure 7.3 that COVI-SCANNER detects a person on the screen (bounding box). As the person moves (steps 1 through 4), the contamination routine highlights the floor as fomite space with several fomite points (red

FIGURE 7.3 Identification of fomite spaces contaminated by infected individuals in the contamination phase.

Containing the Spread of COVID-19 with IoT 137

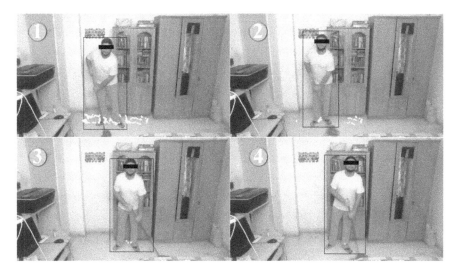

FIGURE 7.4 Removal of the fomite spaces identified in the contamination phase on detection of cleaning activity in sanitization phase.

dots). We comment that the proposed COVI-SCANNER identifies the fomite spaces efficiently.

7.4.2.2 Output from Sanitization Phase

We capture the same instance corresponding to that in Section 7.4.2.1. We observe in Figure 7.4 that the sanitization phase in COVI-SCANNER first identifies an individual (bounding box) and then the broom. Upon detection of the cleaning activity, the COVI-SCANNER starts removing the fomite spaces. At the end of the cleaning activity (steps 1 through 4), we observe that almost all of the fomite spaces are removed as the cleaning object passes through them. We may safely comment that the proposed COVI-SCANNER works efficiently in tracking the sanitization of the detected fomite spaces.

7.4.2.3 Detection and Removal of Fomite Spaces

We perform a cumulative count of the fomite points with each iteration for each of the phases. We observe in Figure 7.5 that the number of fomite spaces increases linearly in the contamination phase. This is because, with each progress in the execution, the contamination routine keeps tracking the infected individual and marks the fomite spaces. On the other hand, we notice that on the detection of the cleaning activity, the sanitization phase removes the corresponding fomite points efficiently. We notice almost 80% removal of the fomite points. We observe a steep decrease in the number of fomite points as the cleaner enters the scene for the first time. This is because the fomite points near the cleaner initially get removed in bulk, compared to the later time instants. Although we observe significant removal of the fomite points, we hardly notice 100% removal of the same. Intuitively, this may be because

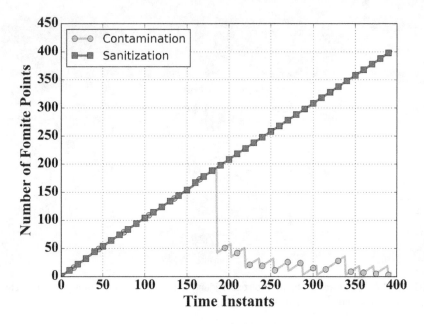

FIGURE 7.5 Number of fomite spaces detected and removed on execution of each phase.

the cleaning object does not pass through all the fomite points or it may be due to the accuracy limitations of the pre-trained models. In the next section, we present the accuracies in each phase.

7.4.2.4 Accuracy of the COVI-SCANNER Phases

We arbitrarily use three videos as inputs to the COVI-SCANNER routine and tabulate the accuracy scores in Table 7.1. We briefly define each of the columns for understanding the results. Accuracy is the quality of correctness of the proposed scheme. Precision represents the correctly identified results and recall represents the number of correctly identified points in comparison to the actual one. The F1-score is the harmonic mean of the precision and recall results, represented as $F1 = 2/(recall^{-1} + precision^{-1})$. On average, we observe an accuracy of 81%. We acquire the precision and recall values to identify the reason behind such low accuracies. We notice that

TABLE 7.1
Accuracy, Precision, Recall, and F1-Score of the COVI-SCANNER Scheme

Videos	Accuracy (%)	Precision (%)	Recall (%)	F1-Score (%)
Video 1	82.75	85.71	80.00	82.75
Video 2	81.08	85.25	76.47	78.7
Video 3	82.05	85.35	77.77	79.99
Mean values	81.96	85.43	78.08	80.51

on average, COVI-SCANNER has a precision of 85% which suggests that it has a fairly high rate of fomite space detection. However, we observe low recall values of 78%, suggesting that the COVI-SCANNER identifies fomite spaces that may not be contaminated. Intuitively, we attribute this behavior to the occlusions that may be present in the video and that of the accuracies of the pre-trained MobileNet-SSD and 3D-ResNet models with 80% and 75%, respectively. In the future, we plan to address this issue and increase the accuracy of the proposed COVI-SCANNER scheme.

7.4.2.5 Delays in Executing COVI-SCANNER and Its Phases

Execution of image processing routines involves complex computations. We account for the high resource demand and propose executing the phases in the COVI-SCANNER scheme periodically after every N frames. Figure 7.6 depicts the time taken in seconds for each frame to be processed in the fog nodes. We observe spikes of almost 1.75 seconds after the N frames due to the delays in acquiring results from the pre-trained activity detection models. As expected, we observe in Figure 7.6 that the frequency of the spikes decreases as we increase N from 10 to 16. This is because of the frequency of the execution of the COVI-SCANNER routines. However, the performance of the COVI-SCANNER starts deteriorating if N is increased as it will miss demarcating most of the fomite spaces, implying the existence of a trade-off between accuracy and the processing delay of COVI-SCANNER. In this chapter, we limit our experiments up to $N = 16$.

We present a granular insight into the delays necessary for executing each phase. In Figure 7.7, we observe that the sanitization phase endures more delay compared to that of the contamination phase. This is because the cleaning activity

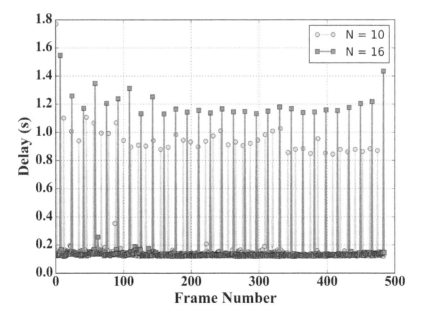

FIGURE 7.6 Delay in processing each frame with varying N.

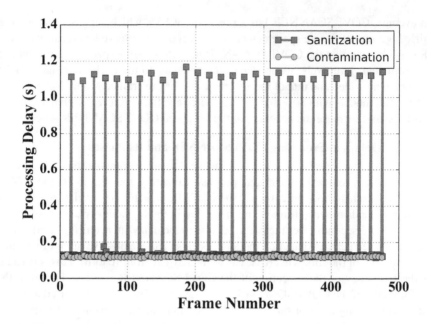

FIGURE 7.7 Delay in processing the frames in the contamination and sanitization phases.

detection model in the sanitization phase requires more time to process as it needs to identify the person, the cleaning objects, and the activity than the person identification and tracking module (contamination). This is because the sanitization phase identifies the person and the object and then the cleaning activity. The contamination phase on the other hand only needs to identify the person and then start tracking.

7.4.2.6 Upload and Download Rates

We capture the upload and download rates using the command line application *iftop* available in Linux distributions. We observe in Figure 7.8 that the contamination phase needs a higher upload rate of almost 800 Kbps, compared to that of the sanitization phase. We observe this variation in the upload rates as the sanitization phase sends only the information about the time instants and coordinates of the fomite points for removal. The contamination phase needs to send the video along with the fomite points while uploading the data for storage at the centralized server. Corresponding to the upload rates from the fog nodes, we observe a maximum download rate of 700 Kbps at the centralized server in Figure 7.9. Such high data rates at the fog nodes and the centralized servers are common because of the large size of videos and images from the cameras, implying the dependency of the proposed COVI-SCANNER on the quality of the network. In the future, we plan to use lightweight streaming protocols to reduce the data rates for each device.

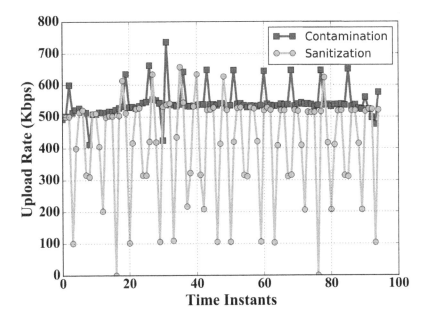

FIGURE 7.8 Upload rate in both contamination and sanitization fog nodes.

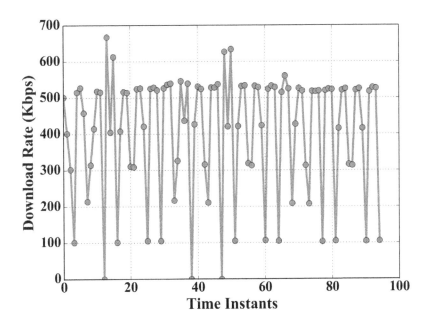

FIGURE 7.9 Download rate at the centralized storage server.

7.5 CONCLUSION

In this chapter, we proposed an AR-based solution named COVI-SCANNER for limiting the secondary spread of the COVID-19 virus in closed public environments. COVI-SCANNER works in two phases: *contamination* and *sanitization*. In the contamination phase, COVI-SCANNER identifies the infected individuals and tracks them to detect fomite spaces and highlight them accordingly. In the sanitization phase, COVI-SCANNER identifies the cleaning objects and the corresponding cleaning activity to remove the fomite points from the contamination phase. In the case of residual fomite points on the screen after the cleaning activity is complete, the centralized server notifies the concerned authorities for re-sanitization. We achieve the identification of the individuals and the concerned activities using pre-trained MobileNet-SSD and 3D-ResNet models. Through deployment and extensive experimentation, we presented the efficiency of the proposed COVI-SCANNER scheme along with the necessary delays and data rates. We hope that with solutions like COVI-SCANNER, we may restrict the secondary spread of the COVID-19 virus in public places and ensure its sanitization.

In the future, we plan to extend this work by increasing the accuracy of detecting and removing the fomite points. Additionally, we also plan to incorporate lightweight streaming protocols to reduce the data rates at the fog nodes and the centralized server.

REFERENCES

Abdellatif, Alaa Awad, Mohamed, Amr, Chiasserini, Carla Fabiana, Tlili, Mounira and Erbad, Aiman. 2019. "Edge Computing for Smart Health: Context-Aware Approaches, Opportunities, and Challenges." *IEEE Network* 196–203.

Barthelemy, Johan, Verstaevel, Nicolas, Forehead, Hugh and Perez, Pascal. 2019. "Edge-Computing Video Analytics for Real-Time Traffic Monitoring in A Smart City." *Sensors* 2048.

Benreguia, Badreddine, Moumen, Hamouma and Merzoug, Mohammed Amine. 2020. "Tracking COVID-19 by Tracking Infectious Trajectories." *arXiv Preprint arXiv:2005.05523*.

Cao, Yu, Chen, Songqing, Hou, Peng and Brown, Donald. 2015. "FAST: A Fog Computing Assisted Distributed Analytics System to Monitor Fall for Stroke Mitigation." *IEEE International Conference on Networking, Architecture and Storage (NAS)* 2–11.

Chamola, Vinay, Hassija, Vikas, Gupta, Vatsal and Guizani, Mohsen. 2020. "A Comprehensive Review of the COVID-19 Pandemic and the Role of IoT, Drones, AI, Blockchain, and 5G in Managing its Impact." *IEEE Access* 90225–90265.

Deb, Pallav Kumar, Misra, Sudip, Mukherjee, Anandarup and Jamalipour, Abbas. 2020. "SkopEdge: A Traffic-Aware Edge-Based Remote Auscultation Monitor." *IEEE International Conference on Communications (ICC)* 1–6.

Gan, Wee Hoe, Lim, John Wah and David, KOH. 2020. "Preventing Intra-Hospital Infection and Transmission of COVID-19 in Healthcare Workers." *Safety and Health at Work*.

Gnouma, Mariem, Ladjailia, Ammar, Ejbali, Ridha and Zaied, Mourad. 2019. "Stacked Sparse Autoencoder and History of Binary Motion Image for Human Activity Recognition." *Multimedia Tools and Applications* 2157–2179.

Hara, Kensho, Kataoka, Hirokatsu and Satoh, Yutaka. 2018. "Can Spatiotemporal 3D CNNs Retrace the History of 2D CNNs and Imagenet?" *IEEE Conference on Computer Vision and Pattern Recognition* 6546–6555.

Howard, Andrew G., Zhu, Menglong and Chen, Bo et al. 2017. "Mobilenets: Efficient Convolutional Neural Networks for Mobile Vision Applications." *arXiv Preprint arXiv:1704.04861.*

Hussain, Adedoyin Ahmed, Bouachir, Ouns, Al-Turjman, Fadi and Aloqaily, Moayad. 2020. "AI Techniques for COVID-19." *IEEE Access* 128776–128795.

Kay, Will, Carreira, Joao and Simonyan, Karen et al. 2017. "The Kinetics Human Action Video Dataset." *arXiv Preprint arXiv:1705.06950.*

Khan, Muhammad Zeeshan, Harous, Saad and Hassan, Saleet Ul et al. 2019. "Deep Unified Model for Face Recognition Based on Convolution Neural Network and Edge Computing." *IEEE Access* 72622–72633.

Khandelwal, Prateek, Khandelwal, Anuj and Agarwal, Snigdha. 2020. "Using Computer Vision to Enhance Safety of Workforce in Manufacturing in a Post COVID World." *arXiv Preprint arXiv:2005.05287.*

Menni, Cristina, Valdes, Ana M. and Freidin, Maxim B. et al. 2020. "Real-Time Tracking of Self-Reported Symptoms To Predict Potential COVID-19." *Nature Medicine* 1–4.

Morales, Giorgio, Salazar-Reque, Itamar, Telles, Joel and Diaz, Daniel. 2019. "Detecting Violent Robberies in CCTV Videos Using Deep Learning." *IFIP International Conference on Artificial Intelligence Applications and Innovations.* Springer. 282–291.

Noori, Farzan Majeed, Wallace, Benedikte, Uddin, Md Zia and Torresen, Jim. 2019. "A Robust Human Activity Recognition Approach Using Openpose, Motion Features, and Deep Recurrent Neural Network." *Scandinavian Conference on Image Analysis.* Springer. 299–310.

Redmon, Joseph and Farhadi, Ali. 2018. "Yolov3: An Incremental Improvement." *arXiv Preprint arXiv:1804.02767.*

Tran, Tuyen X and Pompili, Dario. 2018. "Joint Task Offloading and Resource Allocation for Multi-Server Mobile-Edge Computing Networks." *IEEE Transactions on Vehicular Technology* 856–868.

Waheed, Abdul, Goyal, Muskan and Gupta, Deepak et al. 2020. "Covidgan: Data Augmentation Using Auxiliary Classifier Gan for Improved COVID-19 Detection." *IEEE Access* 91916–91923.

Wang, Bowen, Sun, Yanjing, Duong, Trung Q., Nguyen, Long D. and Hanzo, Lajos. 2020. "Risk-Aware Identification of Highly Suspected COVID-19 Cases in Social IoT: A Joint Graph Theory and Reinforcement Learning Approach." *IEEE Access* 115655–115661.

Wong, Peter Kok-Yiu and Cheng, Jack C.P. 2020. "Monitoring Pedestrian Flow on Campus with Multiple Cameras Using Computer Vision and Deep Learning Techniques." *CIGOS 2019, Innovation for Sustainable Infrastructure* 1149–1154.

Xu, Cheng, Chai, Duo, He, Jie, Zhang, Xiaotong and Duan, Shihong. 2019. "InnoHAR: A Deep Neural Network for Complex Human Activity Recognition." *IEEE Access* 9893–9902.

Younis, Ayesha, Shixin, Li, Jn, Shelembi and Hai, Zhang. 2020. "Real-Time Object Detection Using Pre-Trained Deep Learning Models MobileNet-SSD." *The 6th International Conference on Computing and Data Engineering* 44–48.

8 Crowd-Sourced Centralized Thermal Imaging for Isolation and Quarantine

Sudershan Kumar, Prabuddha Sinha, and Sujata Pal

CONTENTS

8.1	Introduction	145
8.2	Innovative Solutions for Fighting against COVID-19	147
	8.2.1 Challenges and Their Solutions	148
8.3	Objective	148
8.4	Thermal Imaging	149
	8.4.1 Infrared Thermal Scanner	150
	8.4.2 Thermal Camera	150
	8.4.3 Medical Uses of Thermal Imaging	152
	8.4.4 Other Uses of Thermal Imaging	153
8.5	Methodology	153
	8.5.1 Algorithm	153
	8.5.2 Data Updating and Timestamping	156
	8.5.3 Data Synchronization	156
8.6	Conclusion	158
8.7	Future Scope	160
References		161

8.1 INTRODUCTION

The huge advent of the global pandemic arising due to COVID-19 has brought about a drastic change in the entire world. It has spread widely to the whole world with its ever-increasing transmissible nature. It has a dramatically high death rate to go along with it, and this has created a state of panic and fear among the general public. Therefore, we now live in a state of uncertainty and economic turmoil. Because of the constant need to avoid any form of contact and maintaining social distancing, contactless Internet of Things (IoT) devices have come to the forefront of various engineering challenges to address the COVID-19 pandemic.

According to the WHO, COVID-19 is a highly infectious disease which is caused by the recently unearthed coronavirus. The majority of the people who have fallen ill due to the coronavirus usually recuperate without any special treatment and care. It may however cause major issues with people who have co-morbidities, as they do not have adequate immunity to withstand the stress of the symptoms caused by COVID-19. The major symptoms of COVID-19 are:

- Fever
- Dry cough
- Tiredness

Along with these, there are a plethora of minor symptoms such as sore throat, diarrhea, conjunctivitis, headache, rashes, and pains. The virus is mainly passed on through droplets that arise from the mouth or nose of an infected person when they cough or sneeze. These droplets settle on surfaces and can be passed on when they come into contact with the mouth or nose of a non-infected person.

Currently, we are suffering from a new coronavirus which shows similar symptoms to SARS-CoV. It is known as a 2019 novel COrona VIrus Disease19 or COVID-19. The world has previously suffered from an acute respiratory syndrome named SARS coronavirus (SARS-CoV) in 2003, which had quite similar symptoms to COVID-19. At that time, many governments of the world responded by setting up thermal imaging sensors at border checkpoints (airports, seaports, border crossings) to check for one of the key symptoms of SARS, fever, as it has been observed that thermal imaging had emerged as the best screening solution for identification of SARS-CoV (Ng et al., 2004). This was a revolutionary step for monitoring temperature without contact. This saves a lot of time compared to the traditional temperature checking using a mercury thermometer. In conventional temperature-checking, the employee comes in close contact with individuals who have already developed the disease and therefore increases the likelihood of transmissible infection.

Thermal screening has come to the forefront as one of the methods for mass screening of people (Dey et al., 2018). This gives rise to various opportunities to use smart thermal cameras during times of crisis which could help city officials or entire countries to better coordinate the effect of global emergencies like the COVID-19. The best way to reduce the effect of such a pandemic is early detection, which can be increased by mass screening of people in frequently visited public places. Thermal screening has emerged as one of the key strategies employed to identify affected individuals that are crossing borders through various modes of transport like trains, buses, and airplanes. Thermal cameras can be installed in train stations, bus terminuses, and airports so that people who are moving around can be screened continuously and efficiently. After such screening, individuals with COVID-19 can be identified, notified, and effectively held by health officials to quarantine, isolate, and be treated, to avoid a global spike in infection rates. It proved effective, as a fever of 100.5 F (38°C) is a threshold symptom of COVID-19 and is easily identified by thermal scanners. It would be good to screen the important places of a busy city as COVID-19 cases are spreading more in the urban areas.

To this effect, we propose a citywide mass thermal screening system with the help of thermal scanners which can help us to keep track of the localities with people having fever-like symptoms. Therefore those localities can be shut down and the people there can be informed, isolated, and quarantined accordingly by the respective authorities as required. Our proposed system has a very important role in such a situation as the new system directly coordinates with city-level information and captures the cases of each area and stores them in a single database. Due to the automatic filtering of the data, we need less manpower which helps to set up more local screening areas. The high performance or high rate of data sampling would be largely helpful for mass screening on a large scale. It may also be used as a security system later on for surveillance in city areas (Raghavendra et al., 2019; Dey et al., 2018).

Our system helps to send and integrate information in real time. The data will be available for free on sharing platforms like GitHub. This will help the public to take precautions in the symptomatic area where chances of spread of the virus are very high. Also, it is helpful for authorities to take a quick decision on the current situation. Furthermore, it can also be a boon for educational institutes and researchers to view and use the open-sourced data for innovative research.

This chapter is organized as follows: In Section 8.2, we discuss the innovative solutions that are being developed to fight the challenges arising due to the COVID-19 pandemic. In Section 8.3, the existing problems in the systems have been highlighted and the objectives to overcome them have been pinpointed. Section 8.4 is dedicated to explaining the detailed way in which thermal screening is carried out using infrared sensors and scanners. It also highlights the uses of thermal imaging in various industries. Section 8.5 shows us the detailed methodology on how to implement the proposed system with the help of an algorithm with proper data updating and synchronization. Section 8.6 summarizes the entire chapter and provides a detailed comparison of the advantages it holds over existing systems. Section 8.7 highlights the future scope of this technology.

8.2 INNOVATIVE SOLUTIONS FOR FIGHTING AGAINST COVID-19

Various innovative solutions have arisen, which help the frontline workers in healthcare, doctors, guards, police, and security personnel. Some of them are listed below:

- Usage of drones to monitor the various localities strictly under quarantine.
- 3D printed ventilators for COVID-19 patients in the ICU because of the shortage of ventilators.
- Robots helping infected ICU patients with medicines and equipment.
- UV-light emitting disinfection robots with 99.9% kill rate.

Various other novel solutions have been provided by engineers which have helped drastically in the fight against COVID-19. Some of them are: open-sourced ventilator projects, 3D printed masks and face shields, UV light for COVID-19 disinfection, designing of touchless elevator panels, systems for avoiding face touch using necklaces and magnetic rings, contact-free social distance alerting, smart glasses for monitoring and controlling the infection spread, smart wearable thermometer for continuous temperature, IoT-based wearable band to track COVID-19 quarantined patients, COVID-19 intelligent diagnosis and treatment assistant program (NCapp), and smart helmet-based novel COVID-19 detection and diagnosis (Udgata and Suryadevara, 2020).

8.2.1 CHALLENGES AND THEIR SOLUTIONS

A variety of challenges have arisen from different domains; some of them are:

- Social distancing.
- Medical supplies manufacturing.
- Contactless delivery of goods and food items.
- Face mask availability.
- Public transportation.
- Connectivity and monitoring.
- COVID-19 early detection test kits.

There are many challenges that arise due to special needs in the event of an epidemic. One of these is to identify people with COVID-19 and then isolate and quarantine the infected persons. The major symptom of COVID-19 is high temperature; thus thermal scanners are needed in order to check the crowd because contact-based measuring techniques are risky. People are using contactless digital infrared thermometers for checking their body temperature because fever is one of the basic symptoms of the COVID-19 virus. These infrared thermal protectors are being used extensively by the health and public safety authorities because of the social distancing guidelines for avoiding the spread of the COVID-19 virus, which usually spreads through touch.

A challenge arises when examining the temperature of all individuals in a large community. This requires a large number of frontline workers who put their life at risk while doing the work within close proximity of the infected person. Individuals are also at risk while standing in a long queue for checking. The frontline workers check each individual one by one. To minimize this, automatic thermal cameras can be used to measure the temperature of individuals as a group and if a group shows a higher spike in temperature they can be quarantined and further tested for the COVID-19 virus. This data can further be uploaded into an open-source database where the government authorities can access the data and make decisions based on areas, and isolate and seal off the area completely if anybody was found to be affected by COVID-19 in order to minimize the spread of the virus. This can be tabulated through the open-source database where area-wise averages can be calculated for the thermal temperatures and appropriate decisions can be taken.

8.3 OBJECTIVE

Thermal screening is the best way to filter out the people having high temperatures for further test of COVID-19. Currently, most countries (Shaikh et al., 2019) including India are using thermal screening in important areas like airports, railway stations, and some government institutions. It is helpful for the huge population to make them aware of their symptoms. Regardless, the current system has quite a few problems:

1. Thermal screening is done separately in each area such that one institution has no information about the others.
2. Government and public authorities are manually interacting with independent institutions.
3. Delays in real-time information.
4. No direct public information sharing.
5. One-by-one screening using infrared sensors puts public health safety workers at risk.

We have proposed a centralized thermal screening system in which we have the following advantages over the existing system:

1. Real-time data sharing at a single place of all local screening stations.
2. Government and public authorities can directly check data of all places in the city or specific areas of the cities or towns in a single place.
3. In our proposed system, prediction and analysis of the number of cases detected are shared with the public.
4. Thermography using thermal cameras is non-contact and does not require any person to do it contentiously, thus reducing the risk of spreading.

8.4 THERMAL IMAGING

Thermal imaging is a non-contact and rapid system for measuring the infrared radiation that is being emitted from objects with the help of a thermal camera. It has been used in various fields like imaging of internal organs where the organs with different temperatures present themselves as a thermogram, which is like an image with different varying contour lines and shades depicting the temperature of that specific region. The distance from the camera is an important aspect as one needs to be at a certain range for accurate sensing (Sruthi and Sasikala, 2013). If something is not within range, it may not give an accurate temperature. This is a major drawback as we need accurate measurements while following the social distancing norms of COVID-19, and thereby not contaminating the camera.

Besides thermal imaging in the human body, there have been various practical implications for infrared sensing and imaging (Kaplan, 2007). But thermal imaging for analysis of medical conditions is of utmost importance and is still a growing trend. Thermal imagery is used to properly understand and diagnose people with fever, diabetes, breast carcinoma, and dermal problems, which in turn induces specific changes in body temperature that can be detected through the measurement of body surface temperature (Shaikh et al., 2019). The major illnesses that can be diagnosed using thermal imaging are:

- Fever
- Skin cancer

- Diabetes
- Breast cancer

In addition to use in medical imagery, there are fields that can be used by intensive learning algorithms that have become invisible to create thermal images for night vision or heat goggles for various special forces in the military. Infrared thermal scanners and thermal cameras are two major devices used for thermal screening.

8.4.1 Infrared Thermal Scanner

Infrared thermal scanners have become the basic scanning device in nearly all public places during this COVID-19 period. There have been various designs of low-cost non-contact infrared thermometers (Long, 2016). They provide an easy-to-make solution with components found in the market that can be easily made and programmed at home using Arduino. The basic circuitry of an infrared thermal scanner is shown in Figure 8.1.

8.4.2 Thermal Camera

The thermal camera is the core of the proposed system. It is the basic block on which mass thermal screening of the proposed system is carried out. The working principle of a thermal camera is clearly depicted in Figure 8.2. The thermal image of a person

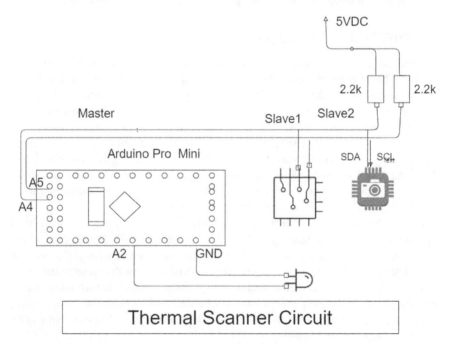

FIGURE 8.1 General circuit of a low-cost infrared thermal scanner.

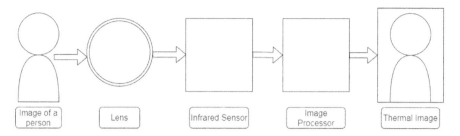

FIGURE 8.2 Working principle of a thermal image camera.

can be captured with the help of infrared sensors which will sense the temperature points in various parts of the lens. These temperature points are then passed through an image processor in order to procure a thermal image that is like a thermal contour with varying temperatures across the image. Various parts of the body depict varying temperatures like the clothed parts are clearly different and can be easily understood from the contours.

There are some factors that affect the performance of the image quality and thermal reading of an object in the camera as mentioned in (FLUKE), which are as follows:

1. *Focus*: The focus of an infrared camera directly affects the accuracy of the temperature measurement data that is captured. An out-of-focus image can produce a temperature measurement that is inaccurate by 20 degrees or more. The main types of focus systems on infrared cameras include fixed focus, manual focus, and automatic focus. Fixed focus cameras are generally designed to scan targets from about 0.45 m (1.5 ft) away or more. Many manual focus cameras can capture images closely.
2. *Optics*: The optics of a thermal image focus on infrared energy onto the detector to produce a response. The materials used for these optics determine how efficiently infrared energy is transmitted to the detector, and therefore the quality of the resulting image.
3. *Spatial resolution*: Both detector pixels and field of view (FOV) play a key role in infrared image details; there is a specification that takes both into account – spatial resolution. The best spatial resolution has the largest number of detector pixels within the smallest field of view.

The field of view (FOV) of a thermal camera is important as the detector resolution. This is used in determining image quality. FOV defines the area the camera sees at a given time. It is determined by the combination of sensor size, lens, and distance to the object. A camera with a wider FOV displays a larger area. One question that arises is how to get an accurate temperature measurement of any size target (for example, 20 mm) and from some distance (say, 15 m) away from the thermal camera. It is important to know both the field of view and the resolution. For this, we have to

FIGURE 8.3 A thermal image after processing.

calculate the instantaneous field of view (IFOV). According to Control Engineering Europe (2018):

$$\text{IF OV (rad)} = (\text{FOV/number of pixels*}) * \left[(3.14/180)(1000)\right] \quad (8.1)$$

$$\text{IF OV (cm)} = (\text{IFOV}/1000) * \text{camera distance to object}, \quad (8.2)$$

Here, the "number of pixels" are indicated as the total number pixels that matches the direction (horizontal/vertical) of the FOV. So, we need to fix the camera at a proper distance in the screening area. Figure 8.3 shows how a thermal image looks after processing (How-To Geek, 2017).

8.4.3 Medical Uses of Thermal Imaging

Infrared thermography is non-contact and non-invasive. Intensive studies of thermal body temperatures using high-definition cameras have been conducted to specifically isolate different parts of the face and oral areas where regression analysis had been done with the temperature for head and canthi (Zhou et al., 2019). These studies have been intensively carried out to screen patients having symptoms of pandemic diseases like SARS, H1N1, and Ebola to stop these outbreaks. It is not possible to screen a person individually for suspected fever in the COVID-19 pandemic. We need to draw inspiration from the technology which had been applied by a group of researchers using infrared thermography through infrared cameras for detecting the fever in suspected SARS patients (Chiu et al., 2005). SARS, just like COVID-19, was a transmittable disease, and it could spread from one person to another through touch. Thus, this was an effective measure to identify SARS patients and specifically isolate them, and put them into specific wards before they come into contact with hospital staff and further transmit the disease. The different types of thermal imaging camera are listed in Figure 8.4.

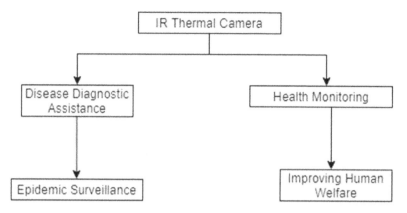

FIGURE 8.4 Medical uses of a thermal imaging camera.

8.4.4 OTHER USES OF THERMAL IMAGING

Other than medical departments, IoT-ready thermal cameras are also found in monitoring temperature constantly at a safe distance in factories and high heat outlets or electrical grids where there is a risk of burn or other hazards (MoviTHERM, 2018). IoT-based thermal security systems have gained popularity over the years and can be used for detection of intruders and trespassers. The intrusion can be detected using infrared sensors along with the Doppler effect (Raghavendra et al., 2019; Wong et al., 2009). Thermal sensors can be used to detect intruding people from the thermal imagery which is portrayed as a hot body with abnormal movements. There are various other uses of the thermal imaging camera (FLIR, Thermal Imaging for Substation Monitoring; FLIR, Thermal Imaging Cameras in the Food Industry) as shown in Figure 8.5.

8.5 METHODOLOGY

In Section 8.4, we discussed the importance of mass screening using thermal cameras. In our proposed system, we extended the existing system such that overall places in the city are monitored efficiently. In our proposed system, we connect local screening systems to the centralized system in the city, such that overall systems monitoring will be done from a single point of control as shown in Figure 8.6.

8.5.1 ALGORITHM

1. When the infrared rays come back to the camera lens after striking the object, the thermal camera reads the image and sends the data to the Raspberry Pi.
2. Raspberry Pi computes the data and sends data using Wi-Fi to the Wi-Fi adaptor.

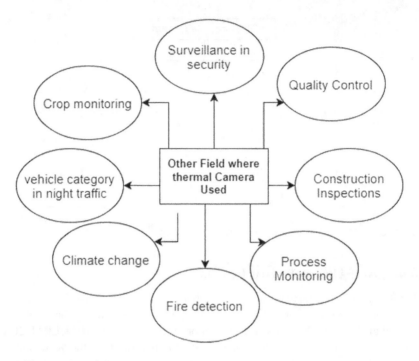

FIGURE 8.5 Uses of the thermal camera.

3. Wi-Fi adaptor sends data to filter and then to the thresholding system which makes an approximation of temperature and filters the data and categorizes it.
4. The local monitor system checks the data value and, if any case is detected, it sends an alarm signal to the screening field.
5. The local system waits for a fixed timestamp and sends data to the centralized system (global) via private access. After this, the data is updated in the global system.
6. The centralized system is unique for a city and contains data of all places in the city.
7. When the data is continuously changing and being updated then the centralized system will be reconfirming the data by sending a request to the local system for checking the system configuration and the update time with a handshaking protocol.
8. Based on the collected information, the system makes important decisions to control the epidemic situation.
9. Data is shared by the public through some open-source media such that the public takes the appropriate precautions.

After filtering value from the local system, the detected value is then sent to the centralized system. The connection between the local and the centralized system is

Crowd-Sourced Centralized Thermal Imaging 155

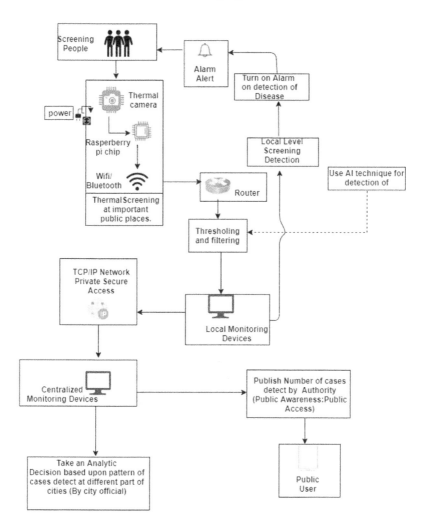

FIGURE 8.6 Proposed system for mass screening at large scale.

secured via a private connection in the proposed system. After the resultant output is generated from the raw data that has been filtered and processed, it is then sent to the city officials for analyzing and checking. They can then make the decisions based on the inferences they have drawn upon. Furthermore, it can be published via an open public network so that it is available to the general public. The government also uses the data to analyze the current situation of the city which helps to strengthen the system accordingly.

We assume that the public user cannot access the system directly. In a centralized system, the related authorities check the pattern. If the patterns show some unusual behavior, then the local authorities will raise the concerns and it can be confirmed by checking the hardware and software system at the local system.

8.5.2 DATA UPDATING AND TIMESTAMPING

A local system updates the value of the database at regular intervals. It detects the number of persons with high temperature, number of screened persons, and previous and current timestamp values. It sends the collected information to the centralized system after every timestamp. The centralized system stores the updated information according to the local system's unique ID. Along with it, the centralized system also stores total cases detected in the city with an updated global or centralized system clock timestamp. The timestamp is used to synchronize the data between the centralized and the local system. The new data is updated after a fixed amount of time which is represented by the difference between the previous timestamp and a new timestamp. When data is passed to the synchronized tool it waits till the next timestamp, after which data is available at the other end. The centralized system concurrently updates a different user ID data and updates global value (total value) with the timestamp. The acknowledgment is sent back to the local system and it confirms the data update. The final information is analyzed and published to the public user to take precautions accordingly. Figure 8.7 depicts how data is shared and timestamped.

8.5.3 DATA SYNCHRONIZATION

For real-time systems, data synchronization is a very important issue. In a real-time system, the data must be consistent having a low error rate and high integrity. Irregularity in data might cause some false inference which may lead to drastic circumstances. Data synchronization updates the data with regular intervals of time and maintains the consistency between two or more devices. It will become challenging when synchronization needs to be done in a remote or mobile network. The main challenge that may occur is the accuracy of data by losing bits in the mobile network, the time interval of syncing of the data security, and data integrity in mobile networks. Data synchronizations mainly deal with updates of the data in all the network nodes and make a single and recently updated copy of the data in the overall system. The real-time analysis helps us to give accurate results in the proposed model.

Many ways of data synchronization have been done in the system. Synchronization is easier when all the systems need synchronization at the same place. However, synchronization becomes complex in different remote places. Clock synchronization (Yuting et al., 2018) is important in many of these protocols and needs some timestamping to order data for event synchronization. There is a local and global clock used to synchronize the data from a local system to a centralized system. Clock timing depends upon several factors such as temperature and clock cycle time. For a real-time distributed system, we have to consider these limitations and apply the best of these algorithms to synchronize the data in the system.

There are two types of clock synchronization, one of which is the external clock synchronization in which time is synchronized based upon the global clock. The second approach is internal clock synchronization in which the network nodes share

Crowd-Sourced Centralized Thermal Imaging

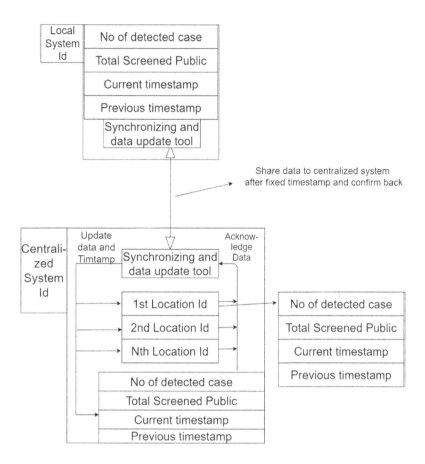

FIGURE 8.7 Data sharing and timestamp.

their timing with other nodes. In our proposal, we need a real-time data update. For this purpose, we need both types of synchronization (Jin and Zeng, 2018). For the real-time system, we need to integrate a relative time synchronization algorithm. They are mainly of two types:

- Cristian's algorithm.
- Berkeley's algorithm.

Cristian's algorithm is a relative time synchronization algorithm. It is mainly a server-client model in which there is only a one-time server. Every time the client node requests the server to synchronize the time. The client node requests periodically and the time server responds with the current UTC (Universal Time Coordinated). Then the client sets its time according to UTC. The client calculates some overhead

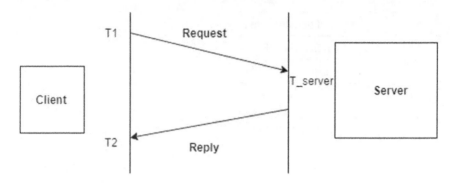

FIGURE 8.8 Cristian's algorithm.

involved during the communication and adjusts the new time (Varma et al., 2013). The new time is:

$$T_{clientnew} = T_{server} + (\text{average over head time}) \tag{8.3}$$

The detailed working of Cristian's algorithm is as shown in Figure 8.8.

Berkeley's algorithm is also a relative time synchronization algorithm. In this algorithm, instead of a single-time server, there are master-slave relationships between the network nodes. Each network node may be used to estimate the time using the Cristian algorithm. The master node sends the request periodically to receive the time of each slave. It responds with the average time and sends the offset of time, and the other nodes adjust their time according to the offset value. Both algorithms accordingly have their limitation and advantages. Figure 8.9 shows the working of Berkeley's algorithm (Varma et al., 2013).

Further, many synchronization algorithms are based upon these two basic algorithms with some modifications such as the Network Time algorithm. Network Time algorithm improves time synchronization over the internet by current UTC based upon the Cristian algorithm. Network Time Protocol and Global Positioning System synchronization are not very efficient for the wireless network because of power constraints. There are various synchronization protocols available for the wireless network; some of them are Reference Broadcast Synchronization (RBS), Timing-Sync Protocol for Sensor Networks (TPSN), and Flooding Time Synchronization Protocol (FTSP) (Sivrikaya and Yener, 2004; Diduch et al., 2008).

8.6 CONCLUSION

Thermal screening has a very important role in the healthcare system during COVID-19. We need an efficient approach to conduct mass thermal screening. In the existing system, the screening is being done at the individual border area of cities such as airports, railway stations, and bus terminuses. Keeping in view the COVID-19 situation, many countries have started to follow mass thermal screening. Unfortunately due to the lack of the smart management of thermal screening, it is not looking very

Crowd-Sourced Centralized Thermal Imaging

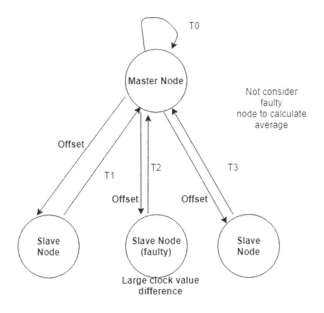

FIGURE 8.9 Berkeley's algorithm.

productive because of delays in the information, misinformation, and improper coordination between the local screening system and the healthcare system. This in turn reduces the efficiency of screening as the proper inferences cannot be carried out by the respective authorities. Due to this problem, there arises a large scope for research in this field in the immediate future.

After SARS-CoV, there has been much advancement in the thermal sensors field but due to the lack of integration with the healthcare system, it is not much used in the healthcare business. Our proposed system of centralized mass thermal screening helps to enhance the healthcare system. It helps to reduce the gap of information in the present system and will automatically update the data of a whole city at a centralized system. It works on the existing system and connects all local screening systems to update real-time information after every timestamp to the centralized system. These pieces of information are helpful in analyzing the future policies of the city's officials. It also informs the public of the affected areas, and thus they can take proper precautions at the correct time while going to the highly affected areas. Quarantine and isolation of respective areas are done in a timely fashion in order to stop the pandemic from spreading further rapidly. One of the major problems which has been addressed in our proposed system is data synchronization. This has been solved using real-time data updates using timestamps. It is important to keep the correct flow of data so that appropriate inferences can be drawn at the correct time.

TABLE 8.1
Existing vs Proposed System

Parameters	Existing System	Proposed System
Monitoring	Only specific places like airports, shopping malls, railway stations, bus terminus and hospitals.	Overall city.
Data Availability	No automatic data sharing between two different places in a city.	Centralized system in the city so that data at a single place is automatically updated.
Analyzing	Due to less coverage, reduced accuracy in estimation of current situation for the pandemic.	High accuracy in estimation of situation due to large coverage.
Responsibility	Specific domain is responsible for monitoring data.	Main authority or government official of the city is responsible for monitoring.

Our proposed system is based upon mass screening considering fever as a symptom. The population size of India is very high. Therefore, we need to set up a local screening area in each city and follow a system as our proposed system will reap significant benefits. Other than the healthcare system, thermal screening is also useful in many areas such as security, quality checking, and many more fields as seen in FLIR, Thermal Imaging Cameras in the Food Industry. It can help in detecting intruders or also can help armed forces with night vision along with various types of medical imagery. It can also help detect anomalies in ceramics and rocks. A detailed comparison is presented in Table 8.1.

As there is no vaccine yet for COVID-19, only medical treatment and care will not save people from the pandemic, but as soon as spread is detected we should be able to isolate the infected people from the mass public. Our system is working on the integration of the existing healthcare system and government organizations with the crowd sourced thermal screening data in entire cities. This proposed system can be further scaled up on a nationwide and worldwide level. The need for integration is due to the unavailability and inaccessibility of the whole city's data in a single place at a specific time. Therefore, our proposed centralized real-time system helps to overcome these difficulties. We can further improve the system by introducing some artificial intelligence (AI) tools along with thermal cameras to detect the symptoms by looking at human behaviors (Tectales, 2020).

8.7 FUTURE SCOPE

In 2003, Severe Acute Respiratory Syndrome (SARS) spread and came in contact with humans in a viral manner. However, it was not widespread and contagious like COVID-19. Few countries such as Singapore, Malaysia, and China which had been affected by SARS then thought to use innovative technologies to overcome the effect of the pandemic in the early stage of the spread. They had used thermal screening and put the thermal cameras in the entry and exit places of the cities (Borsuk, 2003). This helped them to control the spread in the early stage and controlled the situation in those regions effectively. This was not implemented in the majority of the countries at that time as they were unaffected. Even though putting thermal cameras in important places is not very beneficial because it only detects the fever and cough problem in the mass population, the overall system was not integrated into a real-time system, and this did not help in constant clock monitoring.

After the advent of COVID-19, many countries have started to invest in thermal screening and related technologies for the detection of viruses in the early phase of the spread. Along with that, we have to look at the limitation of the current system and work to modernize that gap to make the system better. Our system is mainly doing minor improvements to remove the limitation of the current system. We have to integrate the thermal screening center of the whole city and get the data in real time. This will helps us to alert the places where the chances of transmission are more. This idea can be further scaled up countrywide when the city-wise centralized system is set up successfully. As the whole city database becomes very large and needs the early detection of the data, we can use some pattern matching and AI tools for early estimation of the result. Apart from thermal detection, many thermal cameras are now using AI techniques to detect head, mouth, and hand movements to check for cough-related issues (Tectales, 2020). Therefore, we can also integrate it with our system. It would help more accurate estimation of the effect. In the future, the thermal camera will not just be used for the detection of the disease but also used as a surveillance system. A centralized system helps to control all the activities in one place. Therefore, we have to learn from SARS how some countries used technology to recover in the early phase of the disease. Now we have to work on this so that in future, no pandemic will spread as badly as COVID-19 has done.

Data security is an important issue that should be of concern in the future and we need immediate planning for it. We need to be clear about what type of data needs to be stored in the system and the duration of time to store the data. We also need to look to improve synchronization by regularly replacing the best synchronization techniques and find better techniques in order to maintain the integrity of the data in the system.

REFERENCES

Borsuk, R. 2003. Asia Adopts Thermal Imaging to Spot Travelers with SARS. *The Wall Street Journal*. https://www.wsj.com/articles/SB105148535546154300

Chiu, W. T., P. W. Lin, H. Y. Chiou, et al. 2005. Infrared Thermography to Mass-Screen Suspected SARS Patients with Fever. *Asia Pacific Journal of Public Health* 17, no. 1: 26–28.

Control Engineering Europe. 2018. How Far Can I Measure with a Thermal Imaging Camera? www.controlengeurope.com/article/163635/How-far-can-I-measure-with-athermal-imaging-camera-.aspx

Dey, N., A. S. Ashour, and A. S. Althoupety. 2018. Thermal Imaging in Medical Science. In Management Association, I. (Ed.), *Computer Vision: Concepts, Methodologies, Tools, and Applications*, ed. M. Khosrow-Pour, 1109–1132. IGI Global.

Diduch, L., A. Fillinger, I. Hamchi. et al. 2008. Synchronization of Data Streams in Distributed Realtime Multimodal Signal Processing Environments Using Commodity Hardware. *2008 IEEE International Conference on Multimedia and Expo.* IEEE, no. 1: 1145–1148.

FLIR. Thermal Imaging Cameras in the Food Industry. https://www.flir.com.au/discover/instruments/process-quality/thermal-imaging-cameras-in-the-food-industry/#:~:text=They%20act%20as%20%E2%80%9Csmart%E2%80%9D%20non,positioned%20almost%20anywhere%20as%20needed.

FLIR. Thermal Imaging for Substation Monitoring. https://www.flir.in/discover/instruments/utilities/thermal-imaging-cameras-for-substation-monitoring/#:~:text=Thermal%20imaging%20technology%20can%20improve,installed%20thermal%20imaging%20camera%20systems.

FLUKE. Understanding the Components of Infrared Image Quality. https://www.fluke.com/en/learn/best-practices/test-tools-basics/infrared-cameras/understanding-the-components-of-infrared-image-quality

How-To Geek. 2017. How Does Thermal Imaging Work? https://www.howtogeek.com/294076/how-does-thermal-imaging-work/

Jin, R. and K. Zeng. 2018. Secure Inductive-Coupled Near Field Communication at Physical Layer. *IEEE Transactions on Information Forensics and Security* 13, no. 12: 3078–3093.

Kaplan, H. 2007. *Practical Applications of Infrared Thermal Sensing and Imaging Equipment.* SPIE Press.

Long, G., 2016. Design of a Non-Contact Infrared Thermometer. *International Journal on Smart Sensing & Intelligent Systems* 9, no. 2: 1110–1129.

MoviTHERM, 2018. IoT-Ready Thermal Smart Cameras for North American Market. https://movitherm.com/2018/12/07/thermal-smart-cameras-north-america/

Ng, E. Y., G. J. L. Kawb, and W. M. Chang. 2004. Analysis of IR Thermal Imager for Mass Blind Fever Screening. *Microvascular Research* 68, no. 2: 104–109.

Raghavendra, M., A. M. Bharath, V. Dhananjaya, et al. 2019. IoT Based Thermal Surveillance and Security System. *International Journal of Intelligence in Science and Engineering (IJISE)* 1, no. 2: 93–98.

Shaikh, S., N. Akhter, and R. Manza. 2019. Current Trends in the Application of Thermal Imaging in Medical Condition Analysis. *International Journal of Innovative Technology and Exploring Engineering (IJITEE)* 8, no. 8: 2708–2712.

Sivrikaya, F., and B. Yener. 2004. Time Synchronization in Sensor Networks: a Survey. *IEEE Network* 18, no. 4: 45–50.

Sruthi, S., and M. Sasikala. 2013. A Low Cost Thermal Imaging System for Medical Diagnostic Applications. *IEEE. International Conference on Smart Technologies and Management for Computing, Communication, Controls, Energy Materials (ICSTM)* 1: 621–623.

Tectales. 2020. COVID-19: Deep Learning-Based Cough Recognition. https://tectales.com/ai/covid-19-deep-learning-based-cough-recognition.html

Udgata, S. K., and N. K. Suryadevara. 2020. Advances in Sensor Technology and IoT Framework to Mitigate COVID-19 Challenges. In *Internet of Things and Sensor Network for COVID-19*, ed. Janusz Kacprzyk, 55–82. Springer, Singapore.

Varma, D. A. C., P. K. K. Reddy, and P. Gopinath. 2013. Performance Comparison of Physical Clock Synchronization Algorithms. *International Journal of Engineering Research and Applications* 3, no. 5: 1355–1364.

Wong, W. K., P. N. Tan, C. K. Loo, et al. 2009. An Effective Surveillance System Using Thermal Camera. *2009 International Conference on Signal Acquisition and Processing.* IEEE, no. 1: 13–17.

Yuting, H., X. Yu, L. Kangli, et al. 2018. Research Overview of Clock Synchronization in Wireless Sensor Network. *Journal of Computer Science Applications and Information Technology* 3, no. 1: 1–10.

Zhou, Y., P. Ghassemi, J. Pfefer, et al. 2019. Large-Scale Clinical Study of "Point of Care" Thermal Imaging for Febrile Patient Detection: Towards Optimal Non-Contact Diagnostics in Disease Pandemics (Conference Presentation). *Optical Fibers and Sensors for Medical Diagnostics and Treatment Applications* XIX, no. 10872.

9 Blockchain Technology for Limiting the Impact of Pandemic
Challenges and Prospects

*Suchismita Swain, Oyekola Peter,
Ramasamy Adimuthu, and Kamalakanta Muduli*

CONTENTS

9.1 Introduction: Background and Driving Forces 166
9.2 Problem Statement 167
9.3 Literature Review 168
9.4 Research Methodology 170
    9.4.1 Identification of Research Area 170
    9.4.2 Review Study Scope and Research Conduct 170
    9.4.3 Extraction of Relevant Data 171
9.5 Practical Applications of Blockchain Technology in Combating COVID-19 171
    9.5.1 Prediction and Spread Prevention 171
    9.5.2 Treatment 172
    9.5.3 Direction and Prospects of Blockchain 173
9.6 Blockchain Technology for Revival of Sectors Post COVID-19 174
    9.6.1 Education 175
    9.6.2 Business, Supply Chain, and Logistics 175
    9.6.3 Agriculture 176
    9.6.4 Banking 177
    9.6.5 Manufacturing 178
    9.6.6 Security 178
9.7 Challenges of Implementing Blockchain Technology 178
9.8 Factors Encouraging Adoption of Blockchain Technology 180
9.9 Discussion 181
9.10 Conclusion and Future Scope 181
References 182

9.1 INTRODUCTION: BACKGROUND AND DRIVING FORCES

Blockchain technology was introduced to the world as the brain behind the invention of the digital currency (Bitcoin). In its fundamental form, it is designed to be a networked record database of all digital events where all the transactions that take place therein are first verified by verified participants within the system, which ensures continuous traceability as well as decentralized security. In this system, each block unit is interconnected and expands or increases proportionally to the number of system participants (Corea, 2019; Crosby et al., 2016). According to the World Economic Forum which was held in 2016, it became clearly evident that information technology would play a major role in the fourth industrial revolution, going by the rapid development of interconnected systems as well as arising opportunities to improve already established systems and operational models (Akter et al., 2019).

This trend is already beginning to manifest given the outbreak of the COVID-19 pandemic. Due to the lockdown, measures were imposed in many countries including the shutting down of many facilities and services ranging from schools, places of worships, logistics facilities, etc.; a lot of companies, individuals, and governments have not shifted to operational models to embrace an internet-based service, such that the statistical use of internet-related services since the outbreak has seen an upward trend compared to the normal state prior to the outbreak. This has manifested in the increased use of social media outlets for entertainment, networking, and information sharing; also, there has been an increase in the use of teleconferencing applications such as Zoom, Google Meet, and Google Hangout (Walker et al., 2020). Several studies have already highlighted the application of blockchain technology in a plethora of areas such as healthcare record management, efficient supply chain management, finance, and other related businesses. This migration toward blockchain technology has been made possible due to the fact that it could be applied in solving a lot of specific problems (Hughes et al., 2019). The huge scope of applications of this technology has affected the way some organizations and government organizations work. As of now, despite the adoption by a lot of disconnected start-up businesses, there is yet to be an application of this technology on a large scale, and this is mainly due to intense pushback from the existing systems, despite the benefits that blockchain technology comes with.

In bringing blockchain technology to the context of the current pandemic, there has been an increase in the utilization of information technologies for online classes, meetings, offices, and business operations, etc., with the adoption of these methodologies. It becomes clear that during and after the pandemic, the manner in which many activities are being carried out will definitely not go back to how they were before the pandemic, given the benefits that this new technology comes with. Additionally, given the current state of the pandemic, the threat posed is not unique to the health sector; hence the impact of COVID-19 and the roles of blockchain digital technology are discussed in detail.

The structural architecture of blockchain technology forms a distributed database of a secured crypto algorithm which controls the functionality of each individual block within the link (Hawlitschek et al., 2020) (Figure 9.1).

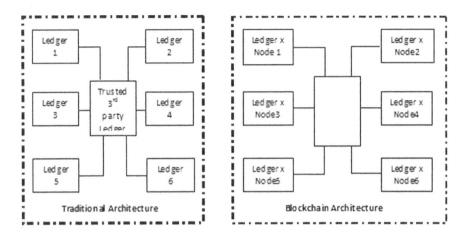

FIGURE 9.1 Traditional and blockchain structural architecture.

Despite the wide applications of blockchain, there are some known challenges to its implementation in a wider scope of application, such as its scalability and performance when it is fused together with other existing systems, as well as the uncertainty of cultural, regional, legal, and even logistical issues which might arise due to its implementation. Despite these, there has been a positive prospect which has prompted a large portion of corporate organizations to adopt blockchain technology. With their success, the global application of this technology is expected to yield similar significant benefits.

The subsequent sections of this study are as follows: Section 9.2 describes the existing problems associated with the use of blockchain technology; Section 9.3 reviews a plethora of related literatures which highlight the global implementation of blockchain technology across diverse fields; Section 9.4 details the method by which relevant articles and literatures were selected as the benchmark for this study; Section 9.5 discusses the practical applications of blockchain technology in government responses to the current COVID-19 situation by analyzing its implementation in prediction and spread prevention, COVID-19 treatment, and its future prospects and direction; Section 9.6 details the numerous ways in which blockchain technology can aid in the economic revival of various sectors after the successful eradication of COVID-19; Section 9.7 also focuses on the challenges of implementation, while Section 9.8 emphasizes the factors encouraging its adoption. Subsequently, Sections 9.9 and 9.10 portray the discussion of results as well as the conclusion of the chapter, together with the future scope.

9.2 PROBLEM STATEMENT

With the continuous increase in the infection rate of the virus, there is a desperate need to develop solutions toward containing the pandemic in a manner that can be easily managed. Technologies such as real-time information and data exchange are essential for developing a reliable vaccine in the shortest possible time. The practical implementation of a technology-based remote patient monitoring system is also essential to safeguard the frontline medical personnel engaged in treating asymptomatic COVID-19 patients. Further, monitoring and tracking exercises are required to furnish throat swabs, personal and contact particulars, and travel details so that health authorities can track and trace those who have made contact with travellers displaying symptoms of COVID-19. This can only be achieved using big data, cloud technology, and Internet of Things (IoT) for the purpose of faster data processing, as well as for determining efficient and effective methodologies for managing the current situation. Additionally, online shopping practices have witnessed a surge of more than 50% in countries like India, Vietnam, and China during the COVID-19 period in a response to enforced pandemic prevention measures such as travel restrictions and social distancing. Many pharma companies are also trying to gain their market share through mobile app-based pharma businesses like Medplus, Apollo pharma, and others. However, widespread adoption of these technologies depends mostly on transparency and data security. Recently, the expansion of cloud-based computing and analytics of big data have seen an increased application in the fight against the pandemic. Nowadays, a massive amount of data could be collected with the application of smart technologies such as mobile phones to gather data in real time. Hence the application of blockchain technology will certainly enhance the quality of digital technology-based services.

9.3 LITERATURE REVIEW

While most research has been previously aimed at the business application of blockchain technology such as in secure transactions (Barnett & Treleaven, 2018), there is now a growing list of possible applications surrounding a vast domain, such as in legal institutions (Halloush & Yaseen, 2019), digital supply chain management (Figorilli et al., 2018; Zhang et al., 2020), energy generation (Aitzhan & Svetinovic, 2018; Erturk et al., 2019), and agriculture (Meroni, 2020). In the field of supply chain management, blockchain technology forms the backbone due to its efficient data processing, efficiency, and transparency as well as reliability of data streams (Bag et al., 2020; Queiroz & Fosso Wamba, 2019). Although most research works have showcased blockchain technology as a universal solution to a variety of problems, there are still questions regarding the fact that there are still some large-scale commercial technologies which already exist. In the work of Iansiti and Lakhani (2017), the authors developed a holistic framework in order to understand the potential of blockchain application within the case organization. From their research, their conclusion supported the impact of this technology given that there were substantial savings arising from operational efficiency, as well as reduction in operational cost. This conclusion was further elaborated in the work of Drescher (Kube, 2018), which was based on the principle of added value that blockchain could offer when compared with the traditionally adopted system. Although there is a significant challenge in adopting blockchain technology in mainstream applications, the fact remains that the security and integrity of blockchain are expensive; hence the commercial reality has to be considered in selecting the most appropriate system to adopt.

More recently, blockchain has been successfully applied in peer-to-peer networks which ensured direct interaction without the need for an intermediary. This structure where there is no need for an intermediary effectively reduced associated cost relating to financial as well as other temporal expenses. In addition to this, the system provides its own storage in the form of a distributed ledger. This stored data offers a safer option with a low possibility of data manipulation. This is achieved by collecting multiple users' data simultaneously, after which the collected data are collectively modified and divided. This ensures data reliability and transparency (Chang et al., 2020; McGhin et al., 2019).

Similarly, in the healthcare sector, blockchain technology has been successfully applied in the management of the COVID-19 pandemic. This is seen in the early detection and identification of outbreaks as well as in research aimed at ensuring more rapid drug delivery, while interconnecting medical systems with the aid of Internet of Things has immensely supported research toward vaccines and possible drugs to counter the effect of the virus (Johnstone, 2020). Furthermore, lots of data are derived from surveillance, forecasting, governmental updates as well as other credible sources (Wong et al., 2019). This could be used to generate an early warning system as well as monitoring trends. From the work of Qin et al. (2020), these data were used to determine the number of active cases of infected people as well as distinguish between whether a case is suspected or confirmed based on social media keywords or search indexes, as well as clinical symptoms. Their adopted method proved to detect suspected and confirmed cases with an estimate of 10 days in advance.

From the agricultural perspective, there has been a transition as well toward the digitalization of processes and structures, either to meet the demand for products or to address concerns relating to sustainability, security, or safety of products. More practically, the use of wireless sensors to monitor agricultural product developments was highlighted in the study conducted by Wang and Li (2013). Similarly, Radio-frequency identification (RFID) technology had been adopted and thoroughly investigated with managerial, operational, and strategic implementation (Vishvakarma et al., 2019), while Heard et al. (2018) studied the application of autonomous vehicles as a means of incorporating sustainable supply chain distribution. Blockchain has also been used in studying the life cycle analysis of agricultural produce impact on the environment (Smetana et al., 2018) This life cycle assessment is usually done in combination with other technologies such as artificial intelligence (AI) and big data (Zhang et al., 2020). In the study conducted by Kukar et al. (2019), a cloud-based system was used to study the behavior and population of agricultural pests as a method to ensure an accurate predictive analysis when making decisions regarding farm management. Other agricultural applications of blockchain technology are studied in more detail in areas such as food traceability (Kamble et al., 2020a) and land registration (Thakur et al., 2020).

As seen in the above-mentioned sectors, blockchain technology has also seen useful and direct application in the power sector even though its full potential is just manifesting itself. The early adopters of blockchain technology have started establishing platforms for digital transactions by capitalizing on the decentralization offered by blockchain. This local adoption ensures that the designs on smart grids can enable local players leverage on the Internet of Things (Erturk et al., 2019). The current system employed is facing a recurrent issue of decreasing revenue with a higher cost of energy; this is mostly due to poorly designed electricity tariffs as well as an unattainable customer payment structure (Aitzhan & Svetinovic, 2018). Significant improvement is possible given the application of blockchain-enabled transactional platforms, which could potentially increase operational efficiency as well as automate a lot of processes. This in turn reduces operational cost.

From here, the subsequent sections will detail the practical applications of blockchain technology in cushioning the negative impact of COVID-19 across a broad range of sectors as well as the challenges and limitations associated with its implementation.

9.4 RESEARCH METHODOLOGY

In the preparation of this chapter, a systematic literature study and mapping was implemented, in order to obtain a broad view of the study area. This was backed up by evidence described in international published research papers which detail the application of blockchain technology. The steps taken for the systematic review process are briefly listed below.

9.4.1 IDENTIFICATION OF RESEARCH AREA

In this step, the objective of the research was to study and understand the role that blockchain technology can play in softening the effect of the pandemic. The primary questions were based on identifying the strength of blockchain technology across various sectors as well as the acceptability in real-world implementations; furthermore, questions regarding its limitation and current challenges faced were considered.

9.4.2 REVIEW STUDY SCOPE AND RESEARCH CONDUCT

In this research stage, a systematic search for existing literatures in the study area was conducted based on data gathered from scientific databases, Web of Science, and Scopus. This ensured that the gathered literatures have been peer-reviewed and subsequently published. Blockchain technology application, blockchain technology in the healthcare sector, COVID-19, and blockchain technology were the search terms used to retrieve papers published between 2014 and 2020. The search retrieved 400 articles. The next step was to eradicate the duplicate papers, which helped us to extract 270 papers, as shown in Figure 9.2. Later it was decided to omit the conference papers, which further reduced the number of papers to 157, and examination of abstracts finally helped us to select 77 papers for our study.

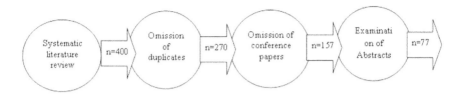

FIGURE 9.2 A systematic review of literature.

9.4.3 EXTRACTION OF RELEVANT DATA

After grouping the literatures into classes, the relevant information was then extracted to address the study questions. Given that the scope of this research cut across various categories, keywords were used to group the selected literature into groups of similar classes based on the contributions of the literature. The subclasses generated for this study are COVID-19, blockchain applications in agriculture, business, energy, finance, health, manufacturing, music, security, and transportation. After grouping the literature into classes, the relevant information was then extracted to address the study questions.

9.5 PRACTICAL APPLICATIONS OF BLOCKCHAIN TECHNOLOGY IN COMBATING COVID-19

Given the broad range of possible applications of blockchain technology as seen in the literature study and drawing from the current issues relating to COVID-19, it is necessary to analyze how this technology can be applied. There have been steady development and use of this system as seen in contact tracing, monitoring and surveillance, research and interconnectivity, and interoperability of clinical facilities in response to the crisis. These direct solutions to the foremost issues are of utmost importance. However, there is still a greater requirement for a critical medical service, proper management of patients and assets, and seamless operation. These extended applications are detailed in the subsection below.

9.5.1 PREDICTION AND SPREAD PREVENTION

A lot of data has been gathered during the ongoing pandemic; some of this data has been obtained from laboratories and medical facilities, immigration, intelligent contact tracing applications, etc., with the combination of blockchain and artificial intelligence. An intelligent analysis of this data could be potentially used in the effective prediction as well as prevention of the spread of the virus.

While blockchain technology covers issues of data security as well as privacy concerns, it can also be used in improving operational efficiency such as tracking an outbreak and supply chain management to allow for efficient distribution of drugs and vaccines (Novikov et al., 2018). Furthermore, the collected data could be used for accurate analysis and pattern recognition, which aids in predictive analysis.

In prevention of virus spread, there are a lot of data visualization solutions currently being used to monitor real-time conditions (Jones, 2020). In this specific application, blockchain technology could be implemented for the verification of data, or used as a storage platform by keeping information relating to location, user information, hotspots zones, etc., in chains of linked blocks which are periodically updated. The final visualized output will therefore represent the current situation of the geographical locations where the technology is being adopted. According to a study by Hao et al. (Xu et al., 2020), blockchain technology was implemented in designing a contact tracing application (BeepTrace) while preserving user privacy, as well as preventing fast battery drain by accumulating the necessary data and storing the data for 14 days, during which COVID symptoms begin to manifest. After this period, the data can be discarded while an address is declared in a 30-minute interval. In a publication by Agam et al. (Bansal et al., 2020), it was suggested that the combination of contact tracing applications with blockchain can be used in the generation of unique identifiers for all user pairs such that whenever contact is made, the summation of transactions is geographically located as well as timestamped. This will enable easy identification of infected persons.

Additionally, there is an increasing requirement for a legal immunity pass in some countries due to COVID-19 (Voo et al., 2020). This pass is intended for use in international travel and workplaces as well as other public areas, and the lack of this might result in a denial of entry. Although this strategy was developed to curtail the spread of the virus, there are however lots of arising issues such as practical and even legal challenges relating to its implementation (Larremore et al., 2020; Vakharia, 2020). These underlying problems could be addressed with the implementation of blockchain technology in preventing the corruption of information as well as monitoring current infection status. Agam et al. (2020) proposed a method in which individuals' registered information were stored in a blockchain network which was used to maintain all COVID-19-related information as well as validate the data collected from individuals and medical facilities to enable seamless data flow. In generating the immunity record, information regarding the antibody test results of the individuals is stored using autonomous keys, and the development of antibodies in an individual is used to determine the immunity period based on the expected maturity age of the antibody. In essence, the time frame in which the antibodies in an individual remain active can then be used to generate a validity period. More interestingly, with the application of blockchain technology, information relating to any individual cannot be extracted or assessed and only validity claims of any individual can be verified using this technique.

9.5.2 Treatment

More recently, a point of care medical approach toward diagnosis and screening have seen increasing adoption in healthcare facilities around the world as a measure to protect medical personnel from coming into contact with an infected patient. This also relieves physical pressure in clinics, which leaves room for more critical cases (Pang et al., 2020; Wang et al., 2020). This method offers rapid screening, which facilitates quick testing of patients. This is critical in reducing fatalities caused by COVID-19 as early detection is now possible. With the application of blockchain as well as artificial intelligence technologies, patients can now test themselves while in isolation from possible exposure.

Tivani and Ellen (Mashamba-Thompson & Crayton, 2020) proposed a blockchain system of self-testing which was based on a mobile application where users could obtain a unique identifier which enables the testing procedure. After the patient is tested, the result could be used to alert relevant bodies of a possible outbreak for a prompt response. With the inbuilt geolocation-enabled feature on the application, there is the ease of positive case tracking which assists in controlled tracing and lockdown decision-making.

Blockchain technology could also be used to facilitate monitoring of patients who are self-isolating in a remote location such as their homes due to the lack of physical space in hospitals. In this scenario, blockchain is combined with other digital technology such as artificial intelligence as well as IoT devices, which are capable of collecting data regarding the patient's health conditions. Blockchain technology can be used as a storage medium for this data, which can then be remotely assessed by authorized medical personnel (Dey et al., 2017; Weiss et al., 2017). In this regard, the use of a smart contracts system on the platform of Ethereum blockchain has been used to manage a remote patient monitoring system (Ali et al., 2020), while blockchain has also been used in the gathering of data and subsequently sharing this information between healthcare facilities (Clim et al., 2019).

9.5.3 Direction and Prospects of Blockchain

Given that blockchain security is based on data scrutiny as all information has to be validated, this feature provides a key solution to the problem of unverified and false information, which is predominant on social media such as Instagram and Facebook, given that technological advancement has made information sharing easier. These tools have also created an easy platform whereby anyone can create fear and panic with "fake news." This forms digital vigilantism (Mirbabaie et al., 2020). Blockchain technology is therefore capable of significantly ensuring that information remains unadulterated and reliable.

Furthermore, there is now a huge demand for potential drugs and vaccines as so many world leaders are doing whatever they can to get their hands on any possible solution to minimize the pandemic. In this regard, medical supply chain comes to play (Gonczol et al., 2020; Sternberg et al., 2020) and the maintenance of steady medical supplies can be made possible with the application of blockchain technology, which helps in continuous movement monitoring of supplies right from their point of origin to their final point of need in a secure manner. This has seen practical application in China (Jones, 2020), where items such as medical gear and masks, using blockchain technology, were quickly dispatched to areas with new outbreaks, which potentially minimized the risk of transmission.

Also, given the huge amount of money donated by private individuals and corporate organizations as well as international aid organizations, it is important to

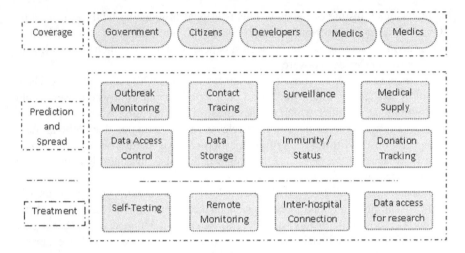

FIGURE 9.3 Blockchain applications in COVID-19 prevention and treatment.

keep track of this financial aid as the use of blockchain in fund management has been widely researched in recent publications (Lukita et al., 2020). As an example, blockchain technology has been used to monitor the donations aimed at supplying face masks to citizens in China. Similarly, other organizations such as health professionals and financial aid organizations can monitor the inflow of funds, as well as track and make sure the funds are utilized as intended by assigning digital signatures and certificates to all donations, which can then be traced through blockchain technology.

Another interesting application of blockchain technology is in the area of medical research, where it is used to prevent the alteration of data as well as give patients the right to allow their medical information to be used for medical research and clinical trials (Figure 9.3). Also, in reviewing COVID-based research works, blockchain technology can be used in the peer-review process as well as a platform for professional education (Funk et al., 2018).

9.6 BLOCKCHAIN TECHNOLOGY FOR REVIVAL OF SECTORS POST COVID-19

The recovery process after the pandemic is over constitutes a critical challenge in rebuilding every nation's economy across all sectors. Mostly, this will require maintenance of financial records, health records, and educational records. This change will also affect social and professional norms. Blockchain technology, as portrayed in the prevailing sections, could be applied in a multitude of sectors ranging from industrial, business, health, etc., while some sectors have seen practical applications; most are on a small scale. Hence, the implications of blockchain technology in economic revival from the pandemic are highlighted below.

9.6.1 Education

The introduction of blockchain technology in the academic field has no shortage of applications even though it has not been fully embraced due to lack of in-depth knowledge on the associated benefits. More importantly, the combination of blockchain with AI and learning analytics has elevated the mode of course delivery as well as management of academic activities.

Blockchain technology provides a platform for secure management of credentials, data, and certificates. Most academic-related activities can also be automated for data sharing as well as validation of certificates (Getso & Johari, 2017) and academic achievements (Arndt & Guercio, 2020; Srivastava et al., 2018). Similarly, Turkanović et al. (2018) proposed an academic grading and credit platform based on the application of blockchain. The system was designed on a peer-to-peer network where the peers were the students and organizations. The students were assigned tokens which were transferred for every completed course based on its unit value, and this enabled every student to globally validate their courses without the need for any intermediary admin staff. The system also includes potential employers in the chain, which makes it easy for organizations to see potential employees based on accumulated skills, grades, or other criteria.

Furthermore, blockchain technology could be used in the validation of intellectual rights ownership such as in patents and creative works. Blockchain could be used to assign a digital identifier with a secured authentication such that the data cannot be modified or unnecessarily accessed, as well as preventing copyright infringement (Wang et al., 2019). Similarly, blockchain technology could be implemented as an intellectual currency, just like Bitcoin, which could be used to purchase anything as long as the merchant accepts that means of payment. In the academic environment, reputation is the backbone of the new digital economy. As such, there is a requirement to build trust through reviews, and this traceability of reputation is already being measured by the citation index or h-index of any research publication. Blockchain technology could be applied in managing many more educational works and records based on other metrics such as university ranking or author rank.

Furthermore, based on this metric system, institutions or tutors could be assigned some form of reputational metric while students could pay for tutoring services using Bitcoin or other forms of blockchain-enabled payment method. Thus, upon passing the course, reputation metrics are exchanged with the student (Sharples & Domingue, 2016). In summary, the educational applications of blockchain technology are in degree management, data storage, certification, access control, fraud reduction, administrative effort reduction and hence lowered operational cost, evaluation of students, and monetization of skills.

9.6.2 Business, Supply Chain, and Logistics

Blockchain technology, which has potential to disrupt existing structures, is a tool which could potentially increase the efficiency and effectiveness of business supply chains. It could be used to keep track of product flow as well as authenticate the sustainability of products, keep automated records of environmental conditions which may be detrimental to service delivery, as well as other health-related concerns (Saberi et al., 2019; Novikov et al., 2018). Furthermore, most systems implementing blockchain in management of supply chains are usually based on a private blockchain network, where only validated users can make changes to the flow of data with a limiting access.

Other business applications include its adaptation in the energy sector as well as management of waste (D. Zhang, 2019) and geographic tracking in supply chains (Sternberg et al., 2020). Also, most customers are concerned about product authenticity, origin, and prior modification, especially when dealing with articles of ostentations such as artwork and jewelry. This detailed information is usually contained in issued certificates or accreditations from third parties to prove genuineness as well as aiding in tracing origin. Past issues relating to unsafe products such as baby formula in China (Qiao et al., 2012) therefore make customers demand for transparent and secured information relating to the product from manufacturers and retailers. Gathering this required information might constitute a roadblock with the increasing complexity in product movements across international boundaries as well as other vast arrays of networks. This means that information relating to products is therefore scattered in multiple places, such that the challenge becomes to aggregate all this information to a unitary location. This is where blockchain plays an important role in data storage and management (Funk et al., 2018).

Logistics services within the business environment branch into subcategories of product handling, storage, and transportation in the traditional scheme. More recently, this definition has been revised such that the purpose of logistics service covers seven main delivery rights: right of product, right of condition, right of time, right of place, right of customer, right of cost, and right of quantity (Queiroz & Fosso Wamba, 2019). These can only be achieved through strategic management of activities relating to the supply of goods. As earlier mentioned, globalization has made it easy for international exchange of goods, and therefore the material flow from the manufacturer to the end consumer becomes complex. This flow of goods is also accompanied by financial flow as well as information flow, which are not often in sync. In exchanging information relating to goods, most processes still involve a fair amount of paperwork, which can account for approximately up to 15–20% of the shipping cost (Lal & Johnson, 2018). The digitalization of these processes will therefore result in an obvious increase in revenue as well as provide a real-time information flow which in turn boosts customer satisfaction.

9.6.3 Agriculture

Just like in the business environment, the agricultural sector is also being pressured to operate in a sustainable manner and comply with best agricultural practices, environment protection, and transparency. These issues become clear given that most agricultural services lack sufficient managerial skills, and there is a general reluctance to embrace technology and industrialization. This leads to fragmentation of relevant information as well as an inefficient supply of produce (Kamble et al., 2020b).

With the application of blockchain technology, artificial intelligence, and IoT, there is a vast increase in the capacity to manage supply and demand as well as adhere to food safety and quality standards. With technology, there is more transparency and accountability across the loop. Maru et al. (2018) developed a data management system which was used to manage purchase records as well as equipment and financial expenses. The advantages of the implementation of this system were a steep decrease in data manipulation and better long-term sustainability. Furthermore, RFID tags were used in blockchain management of food safety (Vishvakarma et al., 2019).

9.6.4 Banking

In the financial sector, there are a lot of associated costs which keep the full system running at capacity; additional costs are assigned toward power and other necessary infrastructures required to maintain employees as well as management of waste. Also, to maintain currency standard, the older currency is always disposed of, shredded, or burned as seen fit. With the establishment of digital currency, the required infrastructures are the internet service without the requirement of power, water, gas, and other paper works required in the traditional financial transaction. The only operational costs in using cryptocurrencies are the transaction validation cost and the cost of distributing, which is a significant reduction in operational cost. This shows the potential of blockchain technology in adaptation to financial activities such as in payment and compliance, as well as it being a tool for foreign currency exchange which can be implemented in businesses for interoperability with other networks (Finextra, 2016).

Theoretically, an interoperable networking system helps to eliminate duplicate record as well as reduce error occurrence. The main concern with this system is the scale at which it can be implemented and the security of data. Furthermore, data processing abilities in terms of speed, reliability, and computational power requirements are some concerns raised (Finextra, 2016).

Due to its potential applications, several banks have adopted this method in an effort to make international transactions in real time (Kube, 2018), due to its projection of a potential 70% reduction in expense given its more optimized operation, which even makes it easier for auditing financial operations (Rajnak & Puschmann, 2020). Furthermore, by using blockchain technology, financial transactions are faster but require validation from data miners. Digital currencies like Bitcoin and Ethereum are capable of validating up to one transaction per minute while Next can process one block every few seconds (Cocco et al., 2017). Recent improvement however has seen an increase in these cryptocurrencies, up to seven transactions per second on average. Further, although traditional payment systems such as VISA and PayPal are capable of handling between 100 and 2,000 transactions per second, the actual time taken between payment and validation is usually between one and four working days, where the variations are caused by difference in currencies or employee factors such as weekends or holidays. When compared with the crypto network, financial settlement averages to 10 minutes irrespective of currency, working days, or holidays (Cocco et al., 2017).

9.6.5 Manufacturing

The extent to which technology meets manufacturing is now at a point where machines can communicate securely with each other to assist in the monitoring of processes and for greater efficiency. However, most business environments remain volatile given the high rate of uncertainty, supply issues, and other operational performance issues. Mostly, the uncertainties may be a result of managerial decisions which can impact the business positively or negatively. By implementing blockchain technology in manufacturing, the storage of data, condition monitoring of equipment, and machine communications could potentially help in reducing the operational cost as well as assist managers in making informed decisions (Rejeb et al., 2019).

Janusz et al. (Sikorski et al., 2017) researched the application of blockchain technology in establishing an electricity market where there were two power-producing companies and a consumer within a blockchain network. The use of blockchain was therefore used by the consumer to analyze the two offers in order to determine the best choice at minimum cost. Another industrial application of this technology is seen in the development of a robust pricing system or in making trade deals by applying smart contracts.

9.6.6 Security

Cyberattacks have become more advanced today, and a lot of companies have fallen victim to security breaches as well as cases of theft of user data (Zhang, 2019). Due to this, a lot of intrusion detectors are being utilized in a plethora of sectors which are based on network monitoring, traffic analysis, etc. A complete security system involves user authentication, which validates the user's identity, user authorization, which verifies the level of access a user has control over in the system, and finally accountability, which reviews system activity.

Blockchain technology which boasts a decentralized system can therefore be applied in maintaining system integrity because it is capable of maintaining data security through the application of cryptography, which can ensure the transmission of data with mutual trust while maintaining user privacy. This system is perfect for applications relating to connected enterprises such as a blockchain of interconnected healthcare (Sotos & Houlding, 2017), medical record access (Shah et al., 2019), banks, industries (Meng et al., 2018), and access control for management of video rights (Fujimura et al., 2015).

9.7 CHALLENGES OF IMPLEMENTING BLOCKCHAIN TECHNOLOGY

The broad application of blockchain technology also comes with some limitations and challenges which might pose a threat to its large-scale applications; some of these hindrances are briefly explained below:

- *Integration with existing systems*: With blockchain technology being a disruptive development, its integration becomes more complex and most organizations are not entirely ready to disrupt business flow and adjust to a new system.
- *Scarcity of skilled blockchain experts*: Given that blockchain is a relatively new area, there is a shortage of skilled personnel that are capable of implementing this technology on a large scale, as it requires a high level of skill and certification.
- *Scalability*: In expanding blockchain applications for us on a wide scale, blockchain becomes tedious to implement. As earlier explained in the transaction rates of the traditional methods, systems such as VISA, Mastercard, and PayPal are capable of processing up to 2,000 transactions per second, although time for transaction validation is lengthy due to currency differences and weekends. Blockchain technologies such as Bitcoin and Ethereum can only handle between 5 and 20 transactions per second, although validation time averages 10 minutes.
- *Blockchain ecosystem*: Given that blockchain boasts a decentralized network, surrounding structures on which the system is to be used must be decentralized as well. Some of these structures to be put in place are decentralized cloud storage, communications system, domains, etc., all of which are still in the developmental stage.
- *Energy*: In the proof of work blockchain mechanism, the process of validating a block requires the solution of a complicated puzzle. Most of the time, this requires the use of advanced computational machines which consume a lot of power. This also limits commercial applications.
- *Data reliability*: In a blockchain network, data are validated before storage. After a transaction has been appended or validated, it becomes impossible to amend the data. Although this ensures data security, there is no reliability and accuracy of the data, given that the data input might be erroneous, and it is impossible to correct.
- *Privacy*: Despite the fact that blockchain's unique selling point is its transparency, not all organizations are pleased with the idea that their data can be viewed by anyone. For instance, in the blockchain application in supply chain management, all participants can view all the data relating to a particular good; although the information cannot be manipulated, this exposes companies' strategies and leverage or may cast a bad light on a particular organization. As such, possibility of participation diminishes.
- *Interoperability*: There are a lot of standalone blockchain systems currently being deployed. However, one major issue remains the fact that these systems cannot be interconnected to work as one due to developmental coding language employed, privacy and security related measures, etc. These variations present a disarrayed system without any possibility of intercommunication; hence adoption on a mass scale becomes problematic. Similarly, lots of organizations would prefer to work on their own blockchain network which adds to the issue of interoperability.

- *Perception*: Given that the background of blockchain technology has its roots in cryptocurrencies such as Bitcoin, which unfortunately comes with a negative image, it becomes difficult for private individuals as well as government agencies to accept this platform. Mostly in the early stage of its development, Bitcoin had quite a reputation as the standard choice of payment for money laundering, fraud, and other criminal dealings such as purchase of weapons and firearms. However, the understanding that blockchain is not all about cryptocurrencies will enable its adoption.
- *Standardization*: Because of the fact that blockchain technology is a decentralized system, there is freedom for developers to design systems which will meet the users' needs; however, this means that there is no standardized format for specific problems, and therefore the problem of restricted communication and collaboration arises.
- *Regulatory clarity*: Given that blockchain ledgers are designed to exist on nodes which are in different locations around the world, this means that the data may be subjected to different state laws and jurisdiction, hence varying in its legal ramifications and applications. Currently the use of blockchain technology has been approved in some countries and states by law; however, future regulations across the globe create uncertainty which discourages mass adoption until there is clarity for the future.

9.8 FACTORS ENCOURAGING ADOPTION OF BLOCKCHAIN TECHNOLOGY

With the potential of blockchain technology to be used in a number of applications, its widespread adoption will be based on its benefits to both organizations and private users. Some of these benefits include:

- *Transparency*: Due to the open nature by which blockchain operates, all data are visible to all parties involved in a transaction, and hence there is complete trust in the system. This is practically beneficial to the healthcare industry.
- *Decentralization*: All stakeholders in the chain can have access to similar records without the need for a governing authority.
- *Data security*: Immutability in blockchain systems ensures that data saved to the system do not get corrupted or modified while protecting the user's identity.
- *Data ownership*: Users in the blockchain have complete rights over their data as well as control of how it is being utilized.
- *Immutability*: Transactions and data cannot be modified after it has been included in the chain. This is because modifying one block in the system would mean that all subsequent blocks will have to be modified as well.
- *Data access*: Blockchain systems comprise of permission systems where there are no restrictions to view data in the public domain as well as private permission-less systems which could be used to limit the access to only predefined users.

- *Efficiency*: Blockchain can handle data transfer between parties and reduces operational costs and administrative waste. Furthermore, the digitalization of operational processes can significantly speed up processing time and enable error reduction.
- *Data availability*: All data stored in a blockchain system is usually duplicated in various nodes and, as such, retrieval of data is guaranteed to prevent data loss or breach.
- *Anonymity*: This remains an important feature for the users in the system. In the blockchain system, the users can only be identified through their unique public key, and a new key could be generated for every transaction.
- *Auditable*: The blockchain system remains a trusted system where data stored cannot be modified. This helps in conducting a system audit, as all the transactions made within the system are visible to all users.

9.9 DISCUSSION

Blockchain technology and its possible areas of application in restoring economical functionalities post-COVID-19 have been explored in preceding sections as a contribution to existing literatures on blockchain technology. Additionally, for better understanding of potential applications, the limitations and benefits such as transparency, security, efficiency, and automation have been analyzed to show why this technology is a viable option that can be applied in nation-building.

In order to get past the challenges associated with its large-scale applications, several researchers have proposed some countermeasures such as storage of data in "off-chains" to solve the issue of scalability (Xia et al., 2017; Gordon & Catalini, 2018) and use of permission systems for better data security and privacy (Iansiti & Lakhani, 2017). Reversal of invalid transactions (Novikov et al., 2018) as well as performance improvement reserves some nodes for only validation and consensus in order to improve performance. Furthermore, in relation to sustainable development, blockchain technology could significantly impact the health sector in the distribution of medications as well as other needed supplies while simultaneously verifying the integrity of medications and foodstuffs. Similarly, logistic movement and supply chains are made easier without the problem of language barriers and cultural interference. Additionally, sectors such as education, finance, and infrastructural development benefit from this technology by ensuring smarter educational systems, management of developmental projects through blockchain, prevention of exploitation due to transparency, fairness, and a more effective management system (Gordon & Catalini, 2018).

9.10 CONCLUSION AND FUTURE SCOPE

This study underlines the impact of blockchain technology application in the recovery phase of the pandemic which will help in nation-building. The characteristic nature of blockchain technology was reviewed as well as its benefits (transparency, trust, security, etc.) and challenges associated with its widespread implementation.

The functionality offered by this technology addresses a broad range of economic sectors which will ultimately lead to value creation, information availability, and reduction of costs associated with administrative tasks.

Additionally, the application of blockchain technology to various economic sectors was discussed to establish its fundamental benefits. This underlying benefit, as well as other considerable future opportunities for better implementation and application, therefore requires that attention should be focused on the application of this emerging technology as all nations struggle to recover from the impact of COVID-19. Moving forward, more research effort should be directed at addressing the challenges associated with blockchain implementation in order to accommodate mass adoption such as transaction rate, minimized system latency, and better security, as well as a commercial system with minimal power requirement. Furthermore, implementation with other promising technologies such as AI, IoT, and big data could potentially maximize the accuracy of systems based on blockchain technology as well as improve analytical capability.

REFERENCES

Aitzhan, N. Z., & Svetinovic, D. 2018. Security and privacy in decentralized energy trading through multi-signatures, blockchain and anonymous messaging streams. *IEEE Transactions on Dependable and Secure Computing* 15(5): 840–852.

Akter, S., Fosso Wamba, S., Barrett, M., & Biswas, K. 2019. How talent capability can shape service analytics capability in the big data environment? *Journal of Strategic Marketing*, 27(6): 521–539.

Ali, M. S., Vecchio, M., Putra, G. D., Kanhere, S. S., & Antonelli, F. (2020). A decentralized peer-to-peer remote health monitoring system. *Sensors*, 20(6): 1656.

Arndt, T., & Guercio, A. 2020. Blockchain-Based transcripts for mobile higher-education. *International Journal of Information and Education Technology* 10(2): 84.

Bag, S., Wood, L. C., Xu, L., Dhamija, P., & Kayikci, Y. 2020. Big data analytics as an operational excellence approach to enhance sustainable supply chain performance. *Resources, Conservation and Recycling* 153: 104559.

Bansal, A., Garg, C., & Padappayil, R. P. 2020. Optimizing the implementation of COVID-19 "Immunity Certificates" using blockchain. *Journal of Medical Systems*, 44(9): 1–2.

Barnett, J., & Treleaven, P. 2018. Algorithmic dispute resolution—the automation of professional dispute resolution using AI and blockchain technologies. *The Computer Journal*, 61(3): 399–408.

Chang, M. C., Hsiao, M. Y., & Boudier-Revéret, M. 2020. Blockchain technology: efficiently managing medical information in the pain management field. *Pain Medicine*, 21(7): 1512–1513.

Clim, A., Zota, R. D., & Constantinescu, R. 2019. Data exchanges based on blockchain in m-Health applications. *Procedia Computer Science*, 160: 281–288.

Cocco, L., Pinna, A., & Marchesi, M. 2017. Banking on blockchain: costs savings thanks to the blockchain technology. *Future Internet*, 9(3): 25.

Corea, F. 2019. The convergence of AI and blockchain. In *Applied Artificial Intelligence: Where AI Can Be Used In Business*, 19–26. Springer, Cham.

Crosby, M., Pattanayak, P., Verma, S., & Kalyanaraman, V. 2016. Blockchain technology: beyond bitcoin. *Applied Innovation* 2: 6–10.

de Oliveira Neto, G. C., Correia, A. D. J. C., & Schroeder, A. M. 2017. Economic and environmental assessment of recycling and reuse of electronic waste: multiple case studies in Brazil and Switzerland. *Resources, Conservation and Recycling*, 127: 42–55.

Dey, T., Jaiswal, S., Sunderkrishnan, S., & Katre, N. 2017, December. HealthSense: a medical use case of internet of things and blockchain. In *2017 International Conference on Intelligent Sustainable Systems (ICISS)*, 486–491. IEEE.

Erturk, E., Lopez, D., & Yu, W. Y. 2019. Benefits and risks of using blockchain in smart energy: a literature review. *Contemporary Management Research*, 15(3): 205–225.

Figorilli, S., Antonucci, F., Costa, C., Pallottino, F., Raso, L., Castiglione, M., ... Sperandio, G. 2018. A blockchain implementation prototype for the electronic open source traceability of wood along the whole supply chain. *Sensors*, 18(9): 3133.

Fujimura, S., Watanabe, H., Nakadaira, A., Yamada, T., Akutsu, A., & Kishigami, J. J. 2015, September. BRIGHT: a concept for a decentralized rights management system based on blockchain. In 2015 IEEE 5th International Conference on Consumer Electronics-Berlin (ICCE-Berlin), 345–346. IEEE.

Funk, E., Riddell, J., Ankel, F., & Cabrera, D. 2018. Blockchain technology: a data framework to improve validity, trust, and accountability of information exchange in health professions education. *Academic Medicine*, 93(12): 1791–1794.

Getso, M. M. A., & Zainudin, J. 2017. "The blockchain revolution and higher education." *International Journal of Information Systems and Engineering* 5(1)57–65. https://doi.org/10.24924/ijise/2017.04/v5.iss1/57.65

Gonczol, P., Katsikouli, P., Herskind, L., & Dragoni, N. 2020. Blockchain implementations and use cases for supply chains-a survey. *IEEE Access*, 8:11856–11871.

Gordon, W. J., & Catalini, C. 2018. Blockchain technology for healthcare: facilitating the transition to patient-driven interoperability. *Computational and Structural Biotechnology Journal*, 16: 224–230.

Halloush, Z. A., & Yaseen, Q. M. 2019, December. A blockchain model for preserving intellectual property. In *Proceedings of the Second International Conference on Data Science, E-Learning and Information Systems*, 1–5.

Hawlitschek, F., Notheisen, B., & Teubner, T. 2020. A 2020 perspective on "The limits of trust-free systems: a literature review on blockchain technology and trust in the sharing economy." *Electronic Commerce Research and Applications*, 40: 100935.

Heard, B. R., Taiebat, M., Xu, M., & Miller, S. A. 2018. Sustainability implications of connected and autonomous vehicles for the food supply chain. *Resources, Conservation and Recycling*, 128: 22–24.

Hughes, L., Dwivedi, Y. K., Misra, S. K., Rana, N. P., Raghavan, V., & Akella, V. 2019. Blockchain research, practice and policy: applications, benefits, limitations, emerging research themes and research agenda. *International Journal of Information Management*, 49: 114–129.

Iansiti, M., & Lakhani, K. R. 2017. *The Truth About Blockchain Harvard Business Review*. Harvard University, hbr.org/2017/01/the-truth-about-blockchain, accessed date: February, 2, 2019.

Johnstone, S. (2020). *A Viral Warning for Change. The Wuhan Coronavirus Versus the Red Cross: Better Solutions Via Blockchain and Artificial Intelligence*. The Wuhan Coronavirus Versus the Red Cross: Better Solutions Via Blockchain and Artificial Intelligence (February 3, 2020).

Jones, D. 2020. Following COVID-19: how the virus is affecting the mobile payments industry. [Retrieved on 26 September from] https://www.mobilepaymentstoday.com/news/following-covid-19-how-the-virus-is-affecting-the-mobile-payments-industry/

Kamble, S. S., Gunasekaran, A., & Gawankar, S. A. 2020a. Achieving sustainable performance in a data-driven agriculture supply chain: a review for research and applications. *International Journal of Production Economics*, 219: 179–194.

Kamble, S. S., Gunasekaran, A., & Sharma, R. 2020b. Modeling the blockchain enabled traceability in agriculture supply chain. *International Journal of Information Management*, 52: 101967.

Kube, N. 2018. *Daniel Drescher: Blockchain Basics: A Non-Technical Introduction in 25 Steps*. Springer, 329–331.

Kukar, M., Vračar, P., Košir, D., Pevec, D., & Bosnić, Z. 2019. AgroDSS: a decision support system for agriculture and farming. *Computers and Electronics in Agriculture*, 161: 260–271.

Lal, R., & Johnson, S. (2018). *Maersk: Betting on Blockchain Background on Maersk. pub.*

Larremore, D. B., Bubar, K. M., & Grad, Y. H. 2020. Implications of test characteristics and population seroprevalence on "immune passport" strategies. *Clinical Infectious Diseases*. https://doi.org/10.1093/cid/ciaa1019

Lukita, C., Hatta, M., Harahap, E. P., & Rahardja, U. 2020. Crowd funding management platform based on block chain technology using smart contracts. *Journal of Advanced Research in Dynamical and Control Systems*. https://doi.org/10.5373/JARDCS/V12I2/S20201236

Maru, A., Berne, D., Beer, J. D., Ballantyne, P. G., Pesce, V., Kalyesubula, S., ... Chavez, J. 2018. *Digital and Data-Driven Agriculture: Harnessing the Power of Data for Smallholders*. Global Forum on Agricultural Research and Innovation. https://doi.org/10.7490/F1000RESEARCH.1115402.1

Mashamba-Thompson, T. P., & Crayton, E. D. 2020. Blockchain and artificial intelligence technology for novel coronavirus disease-19 self-testing. *Diagnostics* 10(4): 198

McGhin, T., Choo, K. K. R., Liu, C. Z., & He, D. 2019. Blockchain in healthcare applications: research challenges and opportunities. *Journal of Network and Computer Applications*, 135, 62–75.

McLean, J. (2016). *Banking on Blockchain: Charting the Progress of Distributed Ledger Technology in Financial Services*. A Finextra White Paper Produced in Associate with IBM.

Meng, W., Tischhauser, E. W., Wang, Q., Wang, Y., & Han, J. 2018. When intrusion detection meets blockchain technology: a review. *IEEE Access*, 6: 10179–10188.

Meroni, G. 2020. Trusted artifact-driven monitoring of business processes using blockchains. In *Modellierung (Companion)*, 45–47.

Mirbabaie, M., Bunker, D., Stieglitz, S., Marx, J., & Ehnis, C. 2020. Social media in times of crisis: learning from hurricane Harvey for the Coronavirus Disease 2019 pandemic response. *Journal of Information Technology*, 35(3):195–213.

Novikov, S. P., Kazakov, O. D., Kulagina, N. A., & Azarenko, N. Y. 2018, September. Blockchain and smart contracts in a decentralized health infrastructure. In *2018 IEEE International Conference "Quality Management, Transport and Information Security, Information Technologies" (IT&QM&IS)*, IEEE, 697–703.

Pang, J., Wang, M. X., Ang, I. Y. H., Tan, S. H. X., Lewis, R. F., Chen, J. I. P., ... Ng, X. Y. 2020. Potential rapid diagnostics, vaccine and therapeutics for 2019 novel coronavirus (2019-nCoV): a systematic review. *Journal of Clinical Medicine*, 9(3): 623–655

Qiao, G., Guo, T., & Klein, K. K. 2012. Melamine and other food safety and health scares in China: comparing households with and without young children. *Food Control* 26(2): 378–386.

Qin, L., Sun, Q., Wang, Y., Wu, K. F., Chen, M., Shia, B. C., & Wu, S. Y. 2020. Prediction of number of cases of 2019 novel coronavirus (COVID-19) using social media search index. *International Journal of Environmental Research and Public Health*, 17(7): 2365–2378.

Queiroz, M. M., & Wamba, S. F. 2019. Blockchain adoption challenges in supply chain: an empirical investigation of the main drivers in India and the USA. *International Journal of Information Management*, 46:70–82.

Rajnak, V., & Puschmann, T. 2020. The impact of blockchain on business models in banking. *Information Systems and e-Business Management*, 1–53. https://doi.org/10.1007/s10257-020-00468-2

Rejeb, A., Keogh, J. G., & Treiblmaier, H. 2019. Leveraging the internet of things and blockchain technology in supply chain management. *Future Internet*, 11(7): 161–182.

Saberi, S., Kouhizadeh, M., Sarkis, J., & Shen, L. 2019. Blockchain technology and its relationships to sustainable supply chain management. *International Journal of Production Research*, 57(7): 2117–2135.

Shah, M., Li, C., Sheng, M., Zhang, Y., & Xing, C. 2019, July. CrowdMed: a blockchain-based approach to consent management for health data sharing. In *International Conference on Smart Health*, Springer, Cham, 345–356.

Sharples, M., & Domingue, J. 2016, September. The blockchain and kudos: a distributed system for educational record, reputation and reward. In *European Conference on Technology Enhanced Learning*, Springer, Cham, 490–496.

Sikorski, J. J., Haughton, J., & Kraft, M. 2017. Blockchain technology in the chemical industry: machine-to-machine electricity market. *Applied Energy*, 195: 234–246.

Smetana, S., Seebold, C., & Heinz, V. 2018. Neural network, blockchain, and modular complex system: the evolution of cyber-physical systems for material flow analysis and life cycle assessment. *Resources, Conservation and Recycling*, 133: 229–230.

Sotos, J., & Houlding, D. 2017. *Blockchains for Data Sharing in Clinical Research: Trust in a Trustless World*. Intel, Santa Clara, CA, Blockchain Appl. Note, 1.

Srivastava, A., Bhattacharya, P., Singh, A., Mathur, A., Prakash, O., & Pradhan, R. 2018, September. A distributed credit transfer educational framework based on blockchain. In *2018 Second International Conference on Advances in Computing, Control and Communication Technology (IAC3T)*, IEEE, 54–59.

Sternberg, H. S., Hofmann, E., & Roeck, D. 2020. The struggle is real: insights from a supply chain blockchain case. *Journal of Business Logistics*, 1–17. https://doi.org/10.1111/jbl.12240

Thakur, V., Doja, M. N., Dwivedi, Y. K., Ahmad, T., & Khadanga, G. 2020. Land records on blockchain for implementation of land titling in India. *International Journal of Information Management*, 52: 101940.

Turkanović, M., Hölbl, M., Košič, K., Heričko, M., & Kamišalić, A. 2018. EduCTX: a blockchain-based higher education credit platform. *IEEE Access*, 6: 5112–5127.

Vakharia, K. 2020. The right to know: ethical implications of antibody testing for healthcare workers and overlooked societal implications. *Journal of Medical Ethics*, 1–3. doi:10.1136/medethics-2020-106467

Vishvakarma, N. K., Singh, R. K., & Sharma, R. R. K. 2019. Cluster and DEMATEL analysis of key RFID implementation factors across different organizational strategies. global. *Business Review*. https://doi.org/10.1177/0972150919847798

Voo, T. C., Clapham, H., & Tam, C. C. 2020. Ethical implementation of immunity passports during the COVID-19 pandemic. *The Journal of Infectious Diseases*, 222(5): 715–718.

Walker, P., Whittaker, C., Watson, O., Baguelin, M., Ainslie, K., Bhatia, S., … Cucunuba Perez, Z. (2020). Report 12: the global impact of COVID-19 and strategies for mitigation and suppression. [Retrived on September 25 from] https://www.imperial.ac.uk/mrc-global-infectious-disease-analysis/covid-19/report-12-global-impact-covid-19/

Wang, C. J., Ng, C. Y., & Brook, R. H. 2020. Response to COVID-19 in Taiwan: big data analytics, new technology, and proactive testing. *Jama*, 323(14): 1341–1342.

Wang, J., Wang, S., Guo, J., Du, Y., Cheng, S., & Li, X. 2019. A summary of research on blockchain in the field of intellectual property. *Procedia Computer Science*, 147: 191–197.

Wang, N., & Li, Z. 2013. Wireless sensor networks (WSNs) in the agricultural and food industries. In *Robotics and Automation in the Food industry*, Woodhead Publishing, 171–199.

Weiss, M., Botha, A., Herselman, M., & Loots, G. 2017, May. Blockchain as an enabler for public health solutions in South Africa. In *2017 IST-Africa Week Conference (IST-Africa)*. IEEE, 1–8.

Wong, Z. S., Zhou, J., & Zhang, Q. 2019. Artificial intelligence for infectious disease big data analytics. *Infection, Disease & Health*, 24(1): 44–48.

Xia, Q., Sifah, E. B., Smahi, A., Amofa, S., & Zhang, X. 2017. BBDS: blockchain-based data sharing for electronic medical records in cloud environments. *Information*, 8(2): 44–59.

Xu, H., Zhang, L., Onireti, O., Fang, Y., Buchanan, W. B., & Imran, M. A. 2020. BeepTrace: blockchain-enabled privacy-preserving contact tracing for COVID-19 pandemic and beyond. *arXiv Preprint arXiv:2005.10103*.

Zhang, D. 2019. Application of blockchain technology in incentivizing efficient use of rural wastes: a case study on Yitong system. *Energy Procedia*, 158: 6707–6714.

Zhang, X., Sun, P., Xu, J., Wang, X., Yu, J., Zhao, Z., & Dong, Y. 2020. Blockchain-based safety management system for the grain supply chain. *IEEE Access*, 8: 36398–36410.

10 A Study on Mathematical and Computational Models in the Context of COVID-19

Dr. Meera Joshi

CONTENTS

10.1	Introduction	188
	10.1.1 Classification of Mathematical Models	188
	10.1.2 Features of Mathematical Models	189
10.2	Study of Mathematical Models for COVID-19	191
	10.2.1 SIR Model	191
10.3	Extensions of the SIR Model	193
	10.3.1 SIR Model with Parameters such as Birth and Death	193
	10.3.2 SIR Model with Vaccine Impact	193
	10.3.3 SIR Model with Impact of Vaccine and Re-Infection Rate	194
10.4	SEIR Model	194
10.5	SUQC Model	195
10.6	Modified SEIR Model for COVID-19	196
10.7	SEIAR (Susceptible–Exposed–Infected–Asymptomatic–Recovered) Model	197
10.8	SEIAR with Hospitalization	198
10.9	Mathematical Model with Rate of Spreading Proportional to Square Root of Time	199
10.10	A Mathematical Model Incorporating Multiple Transmission Pathways Including Environment to Humans	201
10.11	Challenges of Modeling and Forecasting the Spread of COVID-19	203
	10.11.1 Accurate Assessment of Viral Transmission	204
10.12	Models Not Addressing the Exit Strategy	204
	10.12.1 Herd Immunity	204
	10.12.2 Seroprevalence Survey for Transmission Dynamics and Herd Immunity	205
10.13	Heterogeneities in Transmission	205
10.14	Conclusions	206
References		206

10.1 INTRODUCTION

The outburst of the COVID-19 pandemic was reported for the first time in December 2019 in the city of Wuhan in China. The COVID-19 crisis, with no cure for the virus, has affected the whole world. Every facet of human life has changed in a shorter span of time than one could imagine. Challenges are hurled at us every day in aspects of the economy, health, and social sectors of society. There is a critical need to control the spread of the pandemic by taking various measures and decisions. In a severe pandemic such as COVID-19, to make a proper long-term economic and medical plan, it is very important to understand the magnitude and propagation of the disease. It is imperative to develop a scientific process which presents the progression of the pandemic and identifies the spread pattern, in order to make decisions related to corrective measures and to distribute resources in an optimal manner. The only procedure which is very economical, apart from being highly scientific in nature, is constructing a mathematical model consisting of equations which is analogous to reality. Such a model provides an indispensable framework based on facts and figures, leading to a concrete and practical solution for any type of situation, even one such as the present pandemic scenario of COVID-19. Mathematical models are created in order to meet the abstract, precise, and significant needs of a complex problem. The goal of mathematical modeling is to deduce the structural and functional properties related to understanding the issues of handling difficult real-life situations. The core activity involved in building this type of model is making a set of basic assumptions about the process by identifying the vital features which impact the phenomenon, and without causing any modifications to the real situation, so that the actual essence of it is preserved. The construction of the model must be refined and vigorous to achieve an evocative conclusion. Though models have limitations, we can rely upon them, as the predictions that emerge from solving a mathematical model are immensely accurate in terms of precision to real situations.

10.1.1 Classification of Mathematical Models

A mathematical model is a brief comprehension of real situations in life. While studying mathematical models, it is essential to know the classification of the various types of models. Classification of models is required to decide some of the characteristics of their structure. Depending on the type of prediction of outcome, mathematical models are classified as deterministic models, and non-deterministic or stochastic models. In deterministic models, randomness is not considered in the preparation of the mathematical formulation of future circumstances, and the model will generate the same results from given initial conditions every time. A non-deterministic or stochastic model has an inbuilt randomness to it and different outputs can be acquired for the same problem (France and Thornley, 1984). In this chapter, some classical and important types of mathematical frameworks in the context of COVID-19 are studied.

10.1.2 FEATURES OF MATHEMATICAL MODELS

The objectives for construction of a mathematical model must be defined clearly and a complete understanding of the system to be modeled is a prerequisite for the creation of the framework describing the model. The framework of the model is a system of equations which govern the system. These equations are solved to obtain meaningful inferences in terms of providing a solution to the real-life problem. The four important phases of mathematical models are construction, studying, validating, and implementation of the models.

During construction of a mathematical model, the analysis of the system is very crucial. It is the fundamental requirement to examine the classification of the model, in the sense of whether it is deterministic or stochastic, to acquire the desired results. It is essential to analyze the interactions in the system logically for meaningful application of models. If it is not possible to establish relationships from logical analysis, the data has to be acquired in order to fit equations to it.

Further, the obtained equations need to be studied to infer results in the perspective of the actual problem. A model can be studied in two ways: as a quantitative model or as a qualitative model. In the case of COVID-19 both perspectives are important, as the quantitative model helps in attaining clarity on the magnitude of the pandemic's impact in terms of numbers, and the qualitative model helps to answer how and why the disease is spreading. Quantitative behavior is applicable to a case-specific model estimating spread in an area and qualitative behavior is generally common for models like the cause for spread. Therefore, one has to construct a model which can handle both quantitative and qualitative features with the maximum possible accuracy. To confirm the efficiency of the model, it is required to estimate the outcome of the model depending on the variation of parameters in the model, i.e. by performing sensitivity analysis. The solution to equations can be found by analytical method or by numerical method. Numerical methods are applied when analytical methods cannot be used. In the case of numerical methods, computer time has to be monitored, and we need to ensure that the models work by consuming minimum computer time with a negligible compromise on precision.

After studying and ensuring the performance of the model, its validation has to be approved by verifying it with experimental data. During verification of the model, the focus has to be on checking whether the assumptions made while moving from a verbal to a mathematical model are correct. If any assumption is not required, it has to be discarded. The structure of the model must be confirmed and, if necessary, it must be changed to make the model effective. The model must be verified for predicting the results for some data which was not used in estimating the model parameters. The efficiency of the model can be tested based on many aspects but most importantly on the ability of prediction and the requirements for computing.

The final stage of the implementation of the model is based on how accurately it is functioning both qualitatively and quantitatively in comparison to other models. A model can be used depending upon the needs of the user, whether it delivers the task assigned, and its effectiveness in giving the results which the user is interested in as a meaningful and genuine output.

Now we shall proceed further with the study of specific models for COVID-19 and understand the various features of those models. In Section 10.3, a study of various mathematical models for COVID-19 is discussed. In Section 10.3.1, an overview of the basic Susceptible–Infected–Recovered (SIR) model is presented, which is followed by the extensions of the SIR model. To make the study meaningful, in Section 10.3.1, the SIR model is extended by considering aspects like birth rate and death rate, which are important from the point of view of real life. The speed of spread of COVID-19 is alarming and the devastation it is causing to humankind is increasing every day. The most important source of spread of COVID-19 is close interaction between the infected and the susceptible. The only way to control such a spread is by vaccination. A high level of protection can be provided by vaccinating the maximum percentage of population so that persons with weak immunity also get protected. In treating diseases caused by viruses, antibiotics are mostly used. Overusage of antibiotics leads to drug resistance of the virus causing the infection. Vaccines help to control the development of drug resistance; hence it is necessary to study the SIR model with the impact of vaccination, which is discussed in Section 10.3.2. The number of re-infection cases of COVID-19 is very low but the risk of spreading the disease increases if the re-infected person has no symptoms, which is usual in the case of COVID-19. In this context, it is important to study a mathematical model that addresses the impact of vaccines along with re-infection rate, as offered in Section 10.3.3. An imperative fact which one must focus on is that the exposure of a person to an infected person determines the intensity of COVID-19; therefore it is necessary to study a model which takes into account the fraction of the population exposed to an infected person, as offered in the Susceptible-Exposed-Infected-Recovered (SEIR) model in Section 10.4. The outbreak of COVID-19 can be minimized by implementing measures like quarantine. A model is studied in Section 10.5 which takes into account the fraction of quarantined and unquarantined population. A deeper and more useful study is covered in Section 10.6 in the form of a modified SEIR model, by dividing the infected population into two subgroups of isolated (in hospital) and non-isolated cases, along with the recovery rate and death rate for both groups. The most dangerous threat in the spread of COVID-19 is the asymptomatic population. The Susceptible–Exposed–Infected–Asymptomatic–Recovered (SEIAR) model in Section 10.7 studies the impact of asymptomatic carriers in the spread of COVID-19. An extension of the SEIAR model is presented in Section 10.8, which accounts for hospitalization and death rates. It is highly important to understand the relationship between the rate of spreading of COVID-19 and the time to provide timely treatment to infected persons to control the disease. The model described in Section 10.9 derives a relationship between the rate of spreading and time. There is a need to explore the multiple ways in which the spread of COVID-19 happens, and one such factor is pathogen concentration in the environment which contributes to increase in the spread of COVID-19. The model in Section 10.10 studies the spread of COVID-19 by focusing on the indirect transmission of the disease from environment to humans. This study is crucial, keeping in mind that some studies are being conducted presently with a perspective of whether COVID-19 is airborne or not. Towards the end of the chapter, in Section 10.11, a brief account of the challenges in modeling and forecasting the spread of COVID-19 is presented, so that more accurate, effective, and realistic models can be formulated in future.

10.2 STUDY OF MATHEMATICAL MODELS FOR COVID-19

Mathematical models have been used to study epidemic dynamics for a long time and the study of COVID-19 is no exception. We require mathematical models in the context of COVID-19 to have clarity on how the disease is spreading, what are the factors which contribute to spreading the disease at an exceptionally faster rate, and its intensity in a particular region. These models are backing up policy decisions throughout the world in response to COVID-19. Most policymakers are relying on these models while planning processes and procedures to control the spread of COVID-19 and protecting the citizens of their countries without subjecting them to random and risky experiments.

We shall start the study from the simplest model and gradually look at more complex models in the context of COVID-19.

10.2.1 SIR Model

The mathematical model proposed by Kermack and McKendrick is a first model for studying "immunology simulating infectious diseases" which is also known as the SIR model, formed by selecting the first letters of the three important groups – susceptible, infectious and recovered – of the population being studied (Kermack and McKendrick, 1933). "S" represents the division of the population or density of the population which is not infected but can contract infection due to low immunity or co-morbidity, "I" represents the division of population or density of the population who are presently infected or can be taken as the number of active cases. "R" represents the division of the population which has recovered from the disease. The number of people constituting each group changes every day as each person's membership can change from one group to another group, which means that S, I, and R are functions of time. The total population with all the three groups together is constant or it remains the same, i.e. $N = S + I + R$. The number of births and deaths are ignored while constructing the model. To improve understanding of the model, the fraction of the groups in total population N is considered to be:

$$\frac{S}{N} + \frac{I}{N} + \frac{R}{N} = 1 \text{ (as } S + I + R = N\text{) or we write } \frac{S(t)}{N} + \frac{I(t)}{N} + \frac{R(t)}{N} = 1$$

Let the ratios $S(t)/N$, $I(t)/N$, and $R(t)/N$ be taken as $s(t)$, $i(t)$, and $r(t)$, so we have:

$$s(t) + r(t) + r(t) = 1$$

The rate at which the infection spreads is dependent on $s(t)$, the fraction of susceptible individuals, $i(t)$, the fraction of infected individuals, and $r(t)$, the fraction of recovered individuals. To control the spread of the COVID-19 disease, the number of susceptible persons and infected individuals must decrease. The number of susceptible persons can be reduced by vaccination and the increase in the number of infected persons can be reduced by isolation of the infected, following social distancing norms, hand hygiene, and wearing a mask (Abao-Ismail, 2020). The rate of transmission of COVID-19 per contact impacts the spread of the disease. Let β be the rate of transmission of the disease from infected to susceptible individuals. β is always negative as the number of susceptible cases decreases with time. When the value of β is low, the spread is slow. The magnitude of change from susceptible to infected depends on the fraction of susceptible persons $s(t)$, fraction of infected persons $i(t)$, and the rate of transmission β, which can be written as a differential equation as given in Equation 10.1:

$$\frac{ds}{dt} = -\beta s(t) i(t) \tag{10.1}$$

Let γ be the rate at which the individuals are shifted from the infected group to the recovered group. If "n" denotes the average number of days for an infected individual to recover, then we have $\gamma = 1/n$. The value of n varies depending upon the strength of a person's immune system, medication, and environment (Abao-Ismail, 2020). As the number of recovered individuals increases consistently, the value of γ is positive. The magnitude of change from the infected group to the recovered group depends on the fraction of infected persons and the value of γ, we have:

$$\frac{dr}{dt} = \gamma i(t) \tag{10.2}$$

Also, it is clear that for higher positive values of γ, individuals recover very quickly and more people move from the infected group to the recovered group. As seen in Figure 10.1, the SIR model is represented graphically.

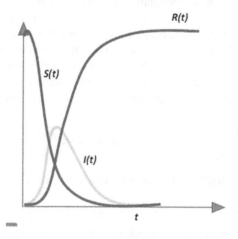

FIGURE 10.1 Graphical representation of the SIR model adapted from Adekola et al. (2020).

Now it is required to focus on the change in the infected group:

$$\frac{di}{dt} = \beta s(t)i(t) - \gamma i(t) \quad (10.3)$$

The rate at which the infection spreads is identified with a measure R_0, which is the average number of people infected by an individual who is already infected, called the basic reproductive number. Many biological and social parameters like incubation period of the virus, host population density, and mode of transmission impact the value of R_0 (Yamamoto, 2018). By letting the total population be 1.0 and the groups a fraction of the total population, R_0 is calculated as $R_0 = \frac{\beta}{\gamma}$. The rate of infection increases if R_0 is positive and decreases when R_0 is negative. It is also used for calculating herd immunity thresholds by the formula (Sasaki, 2020):

$$\text{Herd immunity threshold} = 1 - \frac{1}{R_0}$$

10.3 EXTENSIONS OF THE SIR MODEL

10.3.1 SIR MODEL WITH PARAMETERS SUCH AS BIRTH AND DEATH

By considering factors or parameters like birth and death, the SIR model can be made more accurate and practical (Yamamoto, 2018). Let "b" be the birth rate and "μ" be the death rate. The birth rate increases the cardinality of the susceptible group and the death rate decreases the cardinality. The magnitude of change can be calculated as:

$$\frac{ds}{dt} = b - \mu s(t) - \beta s(t)i(t) \quad (10.4)$$

Death rate decreases the infected population and also it reduces the population in the recovered group and hence is obtained as:

$$\frac{di}{dt} = \beta s(t)i(t) - (\gamma + \mu)i(t) \quad (10.5)$$

$$\frac{dr}{dt} = \gamma i(t) - \mu r(t) \quad (10.6)$$

10.3.2 SIR MODEL WITH VACCINE IMPACT

The impact of vaccines on understanding the spread and control of COVID-19 was studied as a mathematical model (Yamamoto, 2018). In the case of children, let "e" be the inoculation rate of vaccine. By addition of this parameter, the population under the recovered group increases; hence we have the model equations as:

$$\frac{ds}{dt} = b(1-e) - \mu s(t) - \beta s(t)i(t) \qquad (10.7)$$

$$\frac{di}{dt} = \beta s(t)i(t) - (\gamma + \mu)i(t) \qquad (10.8)$$

$$\frac{dr}{dt} = be + \gamma i(t) - \mu r(t) \qquad (10.9)$$

10.3.3 SIR Model with Impact of Vaccine and Re-Infection Rate

The combined effect of vaccine and re-infection rate was studied by Satoru Yamamoto (Yamamoto, 2018). A smaller section of the population might become re-infected at a rate of re-infection ($\beta\sigma$) and the model's equations are:

$$\frac{ds}{dt} = b(1-e) - \mu s(t) - \beta s(t)i(t) \qquad (10.10)$$

$$\frac{di}{dt} = \beta \left[s(t) + \sigma r(t) \right] i(t) - (\gamma + \mu)i(t) \qquad (10.11)$$

$$\frac{dr}{dt} = be + \gamma i(t) - \mu r(t) - \beta \sigma r(t)i(t) \qquad (10.12)$$

10.4 SEIR MODEL

The intensity of disease strongly depends on the exposure of an individual to the symptomatic or asymptomatic carrier of virus. There is a need to add another group of exposed individuals, as this would result in a more effective model. This group is a stage between susceptible and infected, so it is a collection of individuals who have been exposed to the infection but are not infected (Abao-Ismail, 2020; Turchin, 2003). Let $e(t)$ be the fraction of the population which is exposed to infection but is not infected. The rate of change of susceptible group is obtained as:

$$\frac{ds}{dt} = -\beta s(t)i(t) \qquad (10.13)$$

Let δ be the coefficient which gives us the possibility of an exposed person contracting the infection. As every susceptible person is exposed to infection, then the impact of β is positive and, if an exposed person can contract the infection, then the impact of δ is negative. Hence the rate of change of exposed population can be formulated as:

$$\frac{de}{dt} = \beta s(t)i(t) - \delta\, e(t) \qquad (10.14)$$

As there is a possibility of exposed persons contracting infection, the impact of δ on the infected persons is positive, and since there is a possibility of infected persons recovering, then the impact of γ on the rate of infected persons is negative; hence we obtain:

$$\frac{di}{dt} = \delta e(t) - \gamma i(t) \tag{10.15}$$

Similarly, the rate of change of the recovered group can be derived as:

$$\frac{dr}{dt} = \gamma i(t) \tag{10.16}$$

This SEIR model is more realistic compared to the SIR model as every exposed person does not contract the infection immediately on exposure.

10.5 SUQC MODEL

A new practical model to study the spread of COVID-19 has been formulated by Zhao et al. The model considers four groups: susceptible, unquarantined infected, quarantined infected, and confirmed infected cases (Abao-Ismail, 2020). $S(t)$, $U(t)$, $Q(t)$, and $C(t)$ represent the number of susceptible cases, the number of unquarantined infected cases, the number of quarantined cases, and the number of confirmed cases, respectively. The total number of infected cases at an instance of time t is (Zhao and Chen, 2020):

$$I(t) = U(t) + Q(t) + C(t) \tag{10.17}$$

Consider α to be the number of persons infected by an unquarantined person. As the number of susceptible persons decreases, we have the rate of change in the susceptible population defined as:

$$\frac{ds}{dt} = \frac{-\alpha U(t) S(t)}{N} \quad \text{where} \quad N = S(t) + U(t) + Q(t) + C(t) \tag{10.18}$$

Consider γ_1 to be the rate at which the unquarantined persons are quarantined and β to be the rate at which the cases are confirmed. As all the susceptible persons are unquarantined and some of the unquarantined population is going to be quarantined at a rate γ_1, the number of the unquarantined population decreases. So the rate of change of the unquarantined population can be derived as:

$$\frac{dU}{dt} = \frac{\alpha U(t) S(t)}{N} - \gamma_1 U(t) \tag{10.19}$$

Similarly we can find the rate of change in the quarantined infected population and rate of change in confirmed infected population as:

$$\frac{dQ}{dt} = \gamma_1 U(t) - \beta Q(t) \tag{10.20}$$

$$\frac{dC}{dt} = \beta Q(t) \tag{10.21}$$

In this model, there is no recovered group but we observe that cases in the quarantined group are effectively removed from the population temporarily till the results of the test are reported. If the diagnosis is confirmed, then they are removed permanently from the population, thus preventing the spread of the disease.

10.6 MODIFIED SEIR MODEL FOR COVID-19

In the SIR model, the population N is divided into three groups, whereas in SEIR, the population is divided into four groups. The modified SEIR model discussed in this section was formulated and computed by Yao-Yu-Yeo et al. In this model, the total population is divided into six groups (Yeo et al., 2020). S is the susceptible segment of the population and E is the exposed segment of the population. The infected segment is further divided into two subgroups I_H and I_C, where H stands for "Hospital" and C stands for "Community." I_H represents the number of individuals who are infected and isolated and I_C is the number of individuals who are infected but not isolated (unreported cases or overlooked mild symptoms). The individuals in the I_H group do not spread the virus but may contribute to in-hospital spread and individuals in the I_C group spread the virus. Also, if a person is infected, there are two possibilities: either the person recovers or dies. If the outcome is recovery, it is represented by R_H and R_C. If the outcome is death, then it is represented by D_H and D_C. It is assumed that the recovered population attains immunity and is no longer susceptible to infection (Bao et al., 2020; Lu R et al., 2020). Some assumptions are made while formulating the model. For example, travel is not permitted, birth rate and death rate are equal, the in-hospital transmission of infection is negligible, and patients develop immunity after recovering from the infection. It is considered that the total population N is divided into six groups as $N = S + E + I_H + I_C + R_H + R_C$. The differential equations in Equations 10.22 express the rate of change in the various population segments by using parameters and their chosen values (Lauer et al., 2020; Lazzerini and Putoto, 2020; Lin et al., 2020; Linton et al., 2020; Liu et al., 2020):

$$\left.\begin{array}{l}\dfrac{ds}{dt} = -\beta \dfrac{I_C}{N} S \quad \dfrac{dI_H}{dt} = \alpha \rho E - \gamma I_H \quad \dfrac{dR_H}{dt} = \gamma(1-c)I_H \quad \dfrac{dD_H}{dt} = \gamma c I_H \\ \dfrac{dE}{dt} = \beta \dfrac{I_C}{N} S - \alpha E \quad \dfrac{dI_C}{dt} = \alpha(1-\rho)E - \gamma I_C \quad \dfrac{dR_C}{dt} = \gamma(1-C)I_C \quad \dfrac{dD_C}{dt} = \gamma c I_C \end{array}\right\} \tag{10.22}$$

Table 10.1 describes the parameters and their chosen values for the SEIR model from the sources mentioned above.

TABLE 10.1
Parameters for Modified SEIR Model Adapted from Yeo et al., 2020

Parameter	Definition	Value
α^{-1}	Incubation period	Erlang with shape 2 and mean 5.2 days
γ^{-1}	Infectious period	Erlang with shape 2 and mean 3 days
R^1	Basic reproduction rate in phase 1	2.9
R^2	Basic reproduction rate in phase 2	2.5
R^3	Basic reproduction rate in phase 3	1
β	Contact rate per unit time	$R\gamma$
ρ	Proportion of exposed that are pretested	0.1
C	Case fatality rate	0.05

The basic reproduction number is related to β as $\beta = R\gamma$. The analysis is divided into three phases:

- Phase 1: No action taken for controlling the pandemic.
- Phase 2: Measures taken to slow down the spread and prepare for the next phase.
- Phase 3: Lockdown, schools online, and closure of public places.

In phases 1 and 2, R is greater than 1 as it is a necessary condition for an outbreak. The differential equations of this model can be solved numerically by employing fourth-order Runge–Kutta method (Hubbard & West, 1991) with the range $0 \leq t \leq 250$ and value of $h = 1/3$. The initial condition is taken as a single infected person. The model reassures the observation made by a prior study (Li et al., 2020). The model is capable of predicting the occurrence of peaks. The model helps in deciding the measures for containing COVID-19.

10.7 SEIAR (SUSCEPTIBLE–EXPOSED–INFECTED–ASYMPTOMATIC–RECOVERED) MODEL

Understanding the modeling of SEIAR is more important than previous models, as the asymptomatic population plays a critical role in the spread of COVID-19. Individuals who are asymptomatic cause huge damage to the process of controlling the pandemic. We understand that the individuals who do not experience symptoms but are positive in serological tests or blood tests are identified as asymptomatic carriers. The SEIAR model was used to elucidate the transmission dynamics of the swine flu outbreak in 2009 (Adekola et al., 2020). The total population N is divided into five segments: susceptible, exposed, infectious, asymptomatic, and recovered; $N = S(t) + E(t) + I(t) + A(t) + R(t)$. Initially, the total population is susceptible. β is the transmission rate from susceptible to other states. α is mean recovery rate. Other fractions (1–p) proceed to the asymptomatic state during the same time (k), with asymptomatic persons having much less ability to transmit the infection. Let q be the factor which decides the transmissibility in asymptomatic individuals, so $0 < q < 1$. The differential equations that describe this are:

$$\frac{ds}{dt} = -\beta S(1+qA) \tag{10.23}$$

$$\frac{dE}{dt} = \beta S(1+qA) - kE \tag{10.24}$$

$$\frac{dI}{dt} = pkE - \alpha I \tag{10.25}$$

$$\frac{dA}{dt} = (1-p)kE - nA \tag{10.26}$$

$$\frac{dR}{dt} = \alpha I + nA \tag{10.27}$$

$$\frac{dC}{dt} = \alpha I \tag{10.28}$$

Here C denotes the cumulative number of infectived individuals.

10.8 SEIAR WITH HOSPITALIZATION

This model is an extension of the SEIAR model, where a fraction of infected individuals require hospitalization. In this model, the population is classified as $N = S(t) + E(t) + I(t) + A(t) + J(t) + D(t)$, where $J(t)$ and $D(t)$ denote the hospitalized and the dead, respectively. The transmission process in this model is described by the differential equations as follows:

$$\frac{ds}{dt} = \mu N(t) - \frac{\beta S(t)I(t) + J(t) + qA(t)}{N} \tag{10.29}$$

$$\frac{dE}{dt} = \frac{\beta S(t)I(t) + J(t) + qA(t)}{N} - (k+\mu)E(t) \tag{10.30}$$

$$\frac{dA}{dt} = k(1-p)E(t) - (y1+\mu)A(t) \tag{10.31}$$

$$\frac{dI}{dt} = kpE(t) - (\alpha + y1 + \mu)I(t) \tag{10.32}$$

$$\frac{dJ}{dt} = \alpha I(t) - (\delta + y2 + \mu J(t)) \tag{10.33}$$

$$\frac{dR}{dt} = y1(A(t) + I(t)) + y2J(t) - \mu R(t) \tag{10.34}$$

$$\frac{dD}{dt} = \delta J(t) \tag{10.35}$$

$$\frac{dC}{dt} = \alpha I(t) \tag{10.36}$$

Here μ is the rate of birth and natural death. $C(t)$ is the cumulative number of infections. The Spanish flu pandemic data in Geneva was obtained by this model.

10.9 MATHEMATICAL MODEL WITH RATE OF SPREADING PROPORTIONAL TO SQUARE ROOT OF TIME

Let H be the number of persons or individuals who are infected. If P is the total population and H is the number of persons who are infected, then the number of persons who are not infected is $P-H$. The rate of change in H is proportional to the product of number of individuals who are infected and the number of individuals who are not infected (Cakir and Savas, 2020). Hence the mathematical equation is:

$$\frac{dH}{dt} = \mu H(P - H) \tag{10.37}$$

where μ is the constant of proportionality.

This model is revised by Cakir and Savas (2020) in the perspective of the COVID-19 pandemic. The variables and parameters in the model are: "t" is the independent variable, time, which is measured in days; $H(t)$ is the variable dependent on time, expressing the number of people infected at a time t; $\frac{dH}{dt}$ is the rate of spread of disease; and μ is the parameter which represents all the factors influencing the spreading rate.

$H(0) = H_0$ (H at $t=0$) is the initial number of infected individuals and $H(t_1) = H_1$ is the number of infected individuals at $t = t_1$ and the number of susceptible individuals is the population P. For the purpose of building a model which is reliable with regard to the spread of COVID-19 with the progression of time and compatible with the actual data, instead of treating μ as constant μ must be treated as a function of time, and is taken to be $\mu(t) = \frac{K}{2\sqrt{t}}$, where K is a positive proportionality constant. Now the problem can be treated as an initial value problem given by:

$$\frac{dH}{dt} = \mu H(P - H), \text{ with initial conditions: } H(0) = H_0 \text{ and } H(t_1) = H_1$$

Substitute $\mu(t) = \dfrac{K}{2\sqrt{t}}$ in $\dfrac{dH}{dt} = \mu H(P-H)$ to get:

$$\dfrac{dH}{dt} = \dfrac{K}{2\sqrt{t}} H(P-H) \qquad (10.38)$$

Now solve the equation above by separating the variables to get the solution as:

$$\ln H - \ln(P-H) = Pk\sqrt{t} + \ln C \qquad (10.39)$$

i.e. $H(t) = \dfrac{PC}{\left(e^{-Pk\sqrt{t}} + C\right)}$, and by substituting the initial condition $H(0) = H_0$, we get:

$$C = \dfrac{H_0}{P - H_0}. \qquad (10.40)$$

To get the value of k, we substitute $H(t_1) = H_1$ and get:

$$k = \dfrac{1}{P\sqrt{t_1}} \ln \ln \dfrac{H_0(P-H_1)}{H_1(P-H_0)} \qquad (10.41)$$

Finally we get $H(t) = \dfrac{P}{\left(1 + \dfrac{1}{C} e^{-d\sqrt{t}}\right)}, \qquad (10.42)$

where $d = \ln \dfrac{H_0(P-H_1)}{H_1(P-H_0)} \Big/ \sqrt{t_1}$.

To illustrate the results of the model, consider a particular case $P = 100000$. Taking the date 15 March 2020 as $t = 0$, then $H(0) = H_0 = 18$, $t_1 = 11$, i.e. on 26 March 2020, and $H_1 = 3629$.

Now we shall assess the cases for the next 30 days by using this model with respect to the three perspectives by plotting the results on a graph.

Perspective 1: With no precautions and safety measures as the situation was during March, represented in the graph as curve for $\alpha = 1$.

Perspective 2: With 10% increase in precautions, represented in graph as $\alpha = 0.9$.

Perspective 3: With 20% increase in precautions, represented in graph as $\alpha = 0.8$.

The trends in the number of cases based on the precautions taken are graphically represented in Figure 10.2.

Similar analysis for deaths can be carried out, and the trends in the number of deaths based on the precautions taken are graphically represented in Figure 10.3.

This model highlights the importance of precautions for protecting ourselves and society at large. All precautions at the individual level and social level are extremely important to slow down the spread of the pandemic. If these are neglected, the change moves in the negative direction rapidly (Cakir and Savas, 2020). The amount of spread in the disease is exponential in such a time; the precautions have a very valuable role in reducing the spread of COVID-19.

Mathematical and Computational Models on COVID-19

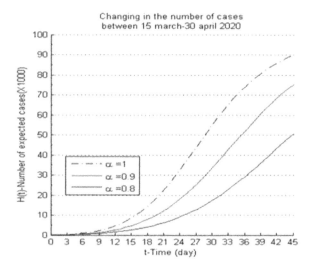

FIGURE 10.2 Graphical representation of trends in the number of cases based on the precautions taken (various values of α represent the curves with increased precautions).

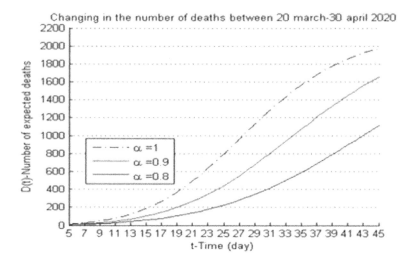

FIGURE 10.3 Graphical representation of trends in the numbers of deaths based on the precautions taken (various values of α represents the curve with increased precautions).

10.10 A MATHEMATICAL MODEL INCORPORATING MULTIPLE TRANSMISSION PATHWAYS INCLUDING ENVIRONMENT TO HUMANS

The complexity in dealing with the infection is caused by several factors which complicate and add challenges in planning measures to control the disease. There is uncertainty or lack of clarity about the origin of the disease, though it is speculated to have originated from wild animals like bats, civets, and minks (Zhou et al., 2020).

Another complication is that infected individuals are asymptomatic for a period of 2–14 days from infection, but they are capable of infecting others at any time (Rothe et al., 2020). All the models studied so far in this chapter, and otherwise also based on the available data, have estimated that the value of basic reproductive numbers for COVID-19 vary from 2.68 to as high as 6.47 (Wu et al., 2020; Tang et al., 2020). A computational model for estimating the extent of the disease outbreak with the center of attention being human-to-human transmission was conducted by Imai et al. (2020). All these models have given prominence to direct human-to-human transmission without accounting for transmission through the environment.

The mathematical model which is being discussed in the present section considers transmission in multiple ways, including transmission from the environment. In this model, the pathogen concentration in the environment is considered, as a susceptible person may contract the infection due to interaction with a contaminated environment. The entire population is divided into four segments as in the SEIR model, i.e. susceptible, exposed, infected, and recovered. Here Λ represents the influx of population, μ is rate of natural death, α is the rate of infection, w is the death rate induced by the disease, γ is the recovery rate from infection, and ξ_1 and ξ_2 are rates of contribution of the virus to environment reservoir from the exposed and infected segments, respectively. V is the concentration of virus in the reservoir of the environment. The transmission rate between susceptible and exposed individuals is represented by the function $\beta_E(E)$. The transmission rate between susceptible and infected individuals is represented by the function $\beta_I(I)$. The indirect transmission rate from environment to human is represented by $\beta_V(V)$. It is assumed that $\beta_I(I)$, $\beta_V(V)$, and $\beta_E(E)$ are increasing functions. By considering the various impacts, which can be either positive or negative, the equations of the mathematical model are framed as follows:

$$\frac{ds}{dt} = \Lambda - \beta_E(E)SE - \beta_I(I)SI - \beta_V(V)SV - \mu S \tag{10.43}$$

$$\frac{dE}{dt} = \beta_E(E)SE + \beta_I(I)SI + \beta_V(V)SV - (\alpha + \mu)E \tag{10.44}$$

$$\frac{dI}{dt} = \alpha E - (w + \gamma + \mu)I \tag{10.45}$$

$$\frac{dR}{dt} = \gamma I - \mu R \tag{10.46}$$

$$\frac{dV}{dt} = \xi_1 E + \xi_2 I - \sigma V \tag{10.47}$$

The above system of equations has a unique disease-free equilibrium at:

$$X_0 = (S_0, E_0, I_0, R_0, V_0) = \left(\frac{\Lambda}{\mu}, 0, 0, 0, 0\right) \tag{10.48}$$

Mathematical and Computational Models on COVID-19

The components which represent infection in this model are E, I, and V. The infection matrix F and the transition matrix V are given by:

$$F = \begin{bmatrix} \beta_E(0)S_0 & \beta_I(0)S_0 & \beta_V(0)S_0 \\ 0 & 0 & 0 \\ 0 & 0 & 0 \end{bmatrix} \quad V = \begin{bmatrix} \alpha+\mu & 0 & 0 \\ -\alpha & w_1 & 0 \\ -\xi 1 & \xi 2 & \sigma \end{bmatrix}$$

where $w_1 = w + \gamma + \mu$. The basic reproduction number of this model is the spectral radius of FV^{-1} which is the next generation matrix (Bender, 1978), i.e. we have:

$$R_0 = \frac{\beta_E(0)S_0}{\alpha+\mu} + \frac{\alpha\beta_I(0)S_0}{w_1(\alpha+\mu)} + \frac{(w_1\xi 1 + \alpha\xi 2)\beta_V(0)S_0}{\sigma w_1(\alpha+\mu)} = R_1 + R_2 + R_3 \quad (10.49)$$

The above equation quantifies the risk due to the disease, where R_1 and R_2 are contributions from human-to-human transmission and R_3 is the contribution from environment-to-human transmission. The overall infection threat due to COVID-19 is from the three means of transmission (Yang and Wang, 2020).

Thus we can estimate the basic reproduction number due to environmental transmission and also due to human-to-human transmission, which helps in planning measures for controlling the spread of COVID-19, more particularly with regard to environmental issues.

Further equilibrium analysis of the above system of equations gives:

$$S = \frac{1}{\mu}\left(\Lambda - (\alpha+\mu)E\right), \; E = \frac{w_1}{\alpha}I, \; R = \frac{\gamma}{\mu}I \text{ and } V = \frac{(w_1\xi 1 + \alpha\xi 2)}{\sigma\alpha}I \quad (10.50)$$

By further analysis we can obtain the pandemic equilibrium of the model.

The study of mathematical models would be incomplete without the discussion of challenges of modeling and forecasting the spread of COVID-19.

10.11 CHALLENGES OF MODELING AND FORECASTING THE SPREAD OF COVID-19

Mathematical modeling has been playing a very important role in guiding policymakers to plan interventions to control the spread of COVID-19. The major aspects which need attention to improve the performance of the mathematical models are:

- It is very important to improve the estimation of parameters in epidemiological models by collecting data from various countries.
- It is essential to focus on the heterogeneities within the population so that the effect of virus transmission and the impact of interventions can be understood clearly to achieve better results in controlling the pandemic.
- There is a need to overcome the shortage of data by collecting more data for clearer understanding and more accurate forecasting.

We will now understand the various issues or challenges in framing mathematical models for COVID-19.

10.11.1 Accurate Assessment of Viral Transmission

In the real-time assessment of transmission of COVID-19, the time-dependent basic reproduction number $R(t)$ is the most important quantity. $R(t)$ represents the number of secondary infections which are caused by an infected individual. If its value is less than 1, then the intensity of the spread of disease will reduce. $R(t)$ is sensitive to the measures implemented for controlling COVID-19. Various mathematical and statistical methods are proposed for estimating the basic reproduction number. Very prominent approaches are the Wallinga–Teunis method (Wallinga and Teunis, 2004) and the Cori method (Cori et al., 2013; Thompson et al., 2019). In spite of having the theoretical framework, some practical issues are to be addressed. Estimation of reproduction number depends on notification data of the cases which might be erroneous due to delay between case onset and the recording of the case. It is crucial to understand how accurately and rapidly the real-time changes in $R(t)$ can be inferred. When the cases are low in number, the accuracy of $R(t)$ decreases. The impact of heterogeneity on the reproduction number needs to be investigated further (Thompson et al., 2020).

10.12 MODELS NOT ADDRESSING THE EXIT STRATEGY

10.12.1 Herd Immunity

It is necessary to plan exit strategies such that relaxations in the interventions for controlling COVID-19 do not affect public health. One of the vital issues in planning an exit strategy is by achieving herd immunity. Herd immunity can be understood as accumulation of a certain level of immunity in the population to prevent further alarming outbreaks. The threshold at which herd immunity is attained is the point at which $R(t)$ drops below 1 for an uncontrolled pandemic. After normalizing the interventions, the possibility of $R(t)$ remaining below 1 depends on the immunity level of the population. If we consider $R(t)$ as deflation of the basic reproduction number R_0, then $R(t)$ can be expressed as:

$$R(t) = \left(1 - i(t)\right)\left(1 - p(t)\right) R_0 \qquad (10.51)$$

where $i(t)$ is the immunity of the community at time t and $p(t)$ represents the reduction factor due to various interventions. After the seizing of all the interventions, the value of $R(t)$ remains less than 1 when $i(t) > 1 - 1/R_0$.

There are some challenges in modeling herd immunity and the threshold of herd immunity. The first challenge we encounter is the lack of clarity on the factors which would influence the threshold of herd immunity, or in some situations the factors are not precisely known. For example, it is not easy to determine immunity level in a community due to reasons like variable sensitivity to antibody tests; also there is ambiguity on whether the individuals with mild symptoms and asymptomatic persons gain immunity. The duration of immunity is not known. Mathematicians need to focus on modeling the threshold of herd immunity inclusive of the aspects mentioned above. The second challenge in calculating herd immunity is an estimation of R_0. There is a lot of variation in the value of R_0 within a country and also between countries. The third challenge is that the spread is not uniform throughout the population (UK Government Office for National Statistics, 2020). It is required to have a proper understanding of various transmission dynamics and to recognize the segment of the population which is most affected to determine the variation in disease-induced immunity and the classical threshold immunity $(1 - 1/R_0)$.

10.12.2 Seroprevalence Survey for Transmission Dynamics and Herd Immunity

Estimation of the fraction of the population which is exposed to the virus directly cannot be detected by regular mechanisms but can be detected through a Seroprevalence survey. This survey helps in tracking COVID-19 among the mild and asymptomatic population which is not detected in laboratory tests. Computational models play a vital role in modeling the data from the Seroprevalence survey. There is a serious need to create mathematical and computational models for addressing these problems which play a key role in planning an exit strategy.

10.13 HETEROGENEITIES IN TRANSMISSION

The modeling of local contact structure with the perspective of spatial heterogeneity and the local mixed structure of households and workplaces is quite successful. But there are some challenges; for example, when the contact networks are altered, it is difficult to track the transmission channels. If a spatial model has a smaller number of near contacts then the number of new infections grows slowly, but with each generation the population of infected persons of the particular generation is larger than the older one, due to which the value of $R(T)$ does not exceed 1 (Riley et al., 2015). Some spatial heterogeneities can be dealt with using clustering and pair approximation models. The vulnerability of an individual to a disease is based on risk factors such as the age of the individual, sex, and background. A model incorporating so many aspects is the target of the scientist. In this situation, a universal challenge faced by any person who is preparing a complex model is the pressure to achieve the targets of the scientific community and make the model simple to attain quick results. If the complexity of the proposed model is high then the implementation of the solution is time-consuming, thus target for a simple model. When more heterogeneities are introduced in epidemiological models, they become complex and bulky in terms of notations and are difficult to solve. In this scenario, it is required to build computational models which can simulate concurrent operations and relations between multiple parameters in an effort to solve complex phenomena. The Agent-Based Model (AGM), which is a category of computational model, plays a vital role in resolving heterogeneity issues. For this, a huge amount of data about human behavior is required. It is important to know human behavior profoundly with regard to their social exposure, their perspective on government regulations, impact of social media, etc. The ABM must be a reflection of our society so that virtual experiments related to policies and measures can be conducted to protect humankind from pandemics like COVID-19.

10.14 CONCLUSIONS

Mathematical models have the potential to understand transmission dynamics, which is the most important feature for controlling a pandemic like COVID-19. The SEIR model is an important extension of the SIR model as it is possible to explain heterogeneous modifications in framework, containment, and analyzing the threats due to transmission of virus. It is observed during the study that mathematical models for COVID-19 have supported policymakers in decision-making through a variety of models, by exploring diverse scenarios to understand parameters such as the transmission of the virus, basic reproductive number, and the importance of a varied range of measures for controlling COVID-19. This chapter will enlighten the reader about the range of models available for understanding and controlling COVID-19. It also gives the reader a vivid explanation of the factors influencing the pandemic, at both micro and macro levels. This study provides a basic idea for the formation of future works in making models according to the real-time situations arising due to the spread of COVID-19. It creates awareness about the nature of COVID-19 and also about the capacity of measures in protecting human health from deadly infections like COVID-19.

The study discusses the challenges faced during building models for COVID-19, as it is an ongoing largest outbreak in the history of the world. It is observed that most of the models are epidemiological models, and there is a need to build more computational models which are based on the concepts of artificial intelligence and machine learning. The novel nature of COVID-19 makes the modeling more challenging. There is an urgent need to collect more data and improve the present models so that they represent the complete real scenario of COVID-19. There is a need to look at the emerging economic challenges due to this outbreak.

REFERENCES

Abao-Ismail, A. 2020. Compartmental Models of the Covid-19 Pandemic for Physicians and Physician-Scientists. *Springer Nature Comprehensive Clinical Medicine*. 2 (May): 852–858. https://doi.org/10.1007/s42399-020-00330-z.

Adekola, H.A., H.O. Egberongbe, I.A. Adekunle, S.A. Onitilo, and I.N. Abdullahi. 2020. Mathematical Modeling for Infectious and Viral Disease: The Covid-19 Perspective. Wiley – *Journal of Public Affairs*. (August). https://doi.org/10.1002/pa.2306.

Bao L,W.Deng, H. Gao, C. Xiao et al. 2020. Lack of Reinfection in Rhesus Macaques Infected With SARS-CoV-2. BioRxiv (May). https://212 doi.https://org/10.1101/2020.03.13.990226.

Bender, E.A. 1978. *An Introduction to Mathematical Modeling*. Dover Publications.

Cakir, Z., and H.B. Savas. 2020. A Mathematical Modeling Approach in the Spread of the Novel 2019 Corona Virus SARS-COV-2(Cocid-19) Pandemic. *Electronic Journal of General Medicine*, 17, no. 4(March): 1–7. https://doi.org/10.29333/ejgm/7861.

Cori, A., N.M. Ferguson, C. Fraser, and S. Cauchemez. 2013. A New Framework and Software to Estimate Time-varying Reproduction Numbers During Epidemics. *American Journal of Epidemiology* 178, no. 9 (September): 1505–1512. https://doi.org/10.1093/aje/kwt133

France, J., and J.H.M. Thornley. 1984. *Mathematical Models in Agriculture.* Butterworths. https://www.thelancet.com/action/showPdf?pii=S0140-6736%2820%2930260-9

Hubbard, J.H. and B.H. West. 1991. *Differential Equations: A Dynamical Systems Approach: Ordinary Differential Equations.* Springer-Verlag. https://doi.org/10.1007/978-1-4612-0937-9.

Imai, N., A. Cori, I. Dorigatti et al. 2020. Report 3: Transmissibility of 2019-nCoV, Reference Source. *Imperial College London COVID-19 Response Team.* https://www.imperial.ac.uk/mrc-global-infectious-disease-analysis/news--wuhan-coronavirus/.

Kermack, W.O., and A.G. McKendrick. 1933. Contributions to the Mathematical Theory of Epidemics III. *Proceedings of Royal Society* 141, no. 843 (July): 94–122. https://doi.org/10.1098/rspa.1933.0106. (re-printed in 1991. Bulletin of Mathematical Biology 53(1/2): 89–118).

Lauer, S.A., K.H. Grantz, Q. Bi, F.K. Jones et al. 2020. The Incubation Period of Coronavirus Disease 2019 (COVID-19) From Publicly Reported Confirmed Cases: Estimation and Application. *Annals of Internal Medicine.* https://doi.org/10.7326/M20-0504.

Lazzerini, M., and G. Putoto. 2020, May. COVID-19 in Italy: Momentous Decisions and Many Uncertainties. *The Lancet* 18. no. 5: E641–E642.

Li, R. et al. 2020. Substantial Undocumented Infection Facilitates the Rapid Dissemination of Novel Coronavirus (SARS-CoV2). *Science.* https://doi.org/10.1126/science.abb3221.

Lin, Q. et al. 2020. A Conceptual Model for the Coronavirus Disease 2019 (COVID-19) Outbreak in Wuhan, China with Individual Reaction and Governmental Action. *International Journal of Infectious Diseases* 93: 211–216. https://doi.org/10.1016/j.ijid.2020.02.058.

Linton, N.M. et al. 2020. Incubation Period and Other Epidemiological Characteristics of 2019 Novel Coronavirus Infections with Right Truncation: A Statistical Analysis of Publicly Available Case Data. *Journal of Clinical Medicine* 9(2): 538. https://doi.org/10.3390/jcm9020538.

Liu, T. et al. 2020. Time-Varying Transmission Dynamics of Novel Coronavirus Pneumonia in China. *BioRxiv, 2020.01.25.919787.* https://doi.org/10.1101/2020.01.25.919787.

Lu, R., X. Zhao, J. Li, P. Niu et al. 2020. Genomic Characterisation and Epidemiology of 2019 Novel Coronavirus: Implications for Virus Origins and Receptor Binding. *The Lancet* 395, no. 10224(February): 565–574. https://doi.org/10.1016/S0140-6736(20)30251-8.

Riley S., K. Eames, V. Isham, D. Mollison, and P. Trapman. 2015. Five Challenges for Spatial Epidemic Models. *Epidemics* 10 (March): 68–71. https://doi.org/10.1016/j.epidem.2014.07.001

Rothe C., M. Schunk, P. Sothmann, G. Bretzel, G. Froeshl, C. Wallrauch et al. 2020. Transmission of 2019-nCoV Infection From an Asymptotic Contact in Germany. *The New England Journal of Medicine* 382, no. 10 (March): 970–971. https://www.nejm.org/doi/10.1056/NEJMc2001468

Sasaki, K. 2020. Covid-19 Dynamics With SIR Model. *The First Cry of Atom.* (March). https://www.lewuathe.com/covid-19-dynamics-with-sir-model.html.

Tang, B., X. Wang, Q. Li, N.L. Bragazzi, S. Tang, Y. Xiao, J. Wu. 2020. Estimation of the Transmission Risk of 2019-nCoV and its Implication for Public Health Interventions. *Journal of Clinical Medicine* 9, no. 2 (February). https://www.mdpi.com/2077-0383/9/2/462.

Thompson, R.N., J.E. Stockwin, R.D. Van Gaalen et al.2019. Improved Inference of Time-varying Reproduction Numbers During Infectious Disease Outbreaks. *Epidemics* 29, no. 100356 (December): 131–145. https://doi.org/10.1016/j.epidem.2019.100356

Thompson, R.N., T.D. Hollingsworth, V. Isham, D. Arribas-Bel et al. 2020. Key Questions for Modeling Covid-19 Exit Strategies. *Proceeding of the Royal Society London B* 287: 20201405. http://dx.doi.org/10.1098/rspb.2020.1405.

Turchin, P. 2003. *Complex Population Dynamics.* Princeton University Press.

UK Government Office for National Statistics. 2020 *Coronavirus (COVID-19) Infection Surveypilot:England,21May(2020).* University of Oxford, the University of Manchester. Public Health England and Wellcome Trust. https://www.ons.gov.uk/peoplepopulationandcommunity/healthandsocialcare/conditionsanddiseases/bulletins/coronaviruscovid19infectionsurveypilot/england21may2020.

Wallinga, J., and P. Teunis. 2004. Different Epidemic Curves for Severe Acute Respiratory Syndrome Reveal Similar Impacts of Control Measures. *American Journal of Epidemiology* 160, no. 6 (September): 509–516. https://academic.oup.com/aje/article/160/6/509/79472.

Wu, T.J., K. Leung, G.M Leung. 2020.Now Casting and Forecasting the Potential Domestic and International Spread of the 2019-nCoV Outbreak Originating in Wuhan, China: A Modelling Study. *The Lancet* 395, no. 10225 (February): 689–697.

Yamamoto, S. 2018. *Introduction to Mathematical Modeling and Computation.* Online Lecture Note. Tohoku University.

Yang, C., and J.Wang. 2020. A Mathematical Model for the Novel Corona Virus Epidemic in Wuhan, China. *AIMS-Mathematical Biosciences and Engineering* 17, no. 3(March): 2708–2724. http://www.aimspress.com/article/10.3934/mbe.2020148.

Yeo, Y.Y., Y.R. Yeo, and W.J. Yeo. 2020. *A Computational Model for Estimating the Progression of Covid-19 Cases in US West and East Coast Regions.* Cambridge University Press. (August): 1–9. https://doi.org/10.1017/exp.2020.45.

Zhao, S., and H. Chen. 2020. Modeling the Epidemic Dynamics and Control of Covid-19 Outbreak in China. Quantitative Biology. (March): 1–9. https://doi.org/10.1007/s40484-20-0199-0.

Zhou, P., X.L. Yang, X.G. Wang et al. 2020. Discovery of Novel Coronavirus Associated With the Recent Pneumonia Outbreak in Humans and its Potential Bat Origin. *bioRxiv.* Cold Spring Harbor Laboratory. (January): 1–18. https://doi.org/10.1101/2020.01.22.914952

11 A Detailed Study on AI-Based Diagnosis of Novel Coronavirus from Radiograph Images

Malaya Kumar Nath and Aniruddha Kanhe

CONTENTS

11.1 Introduction 209
11.2 COVID-19 Etiology, Clinical Imaging Features, and Prognosis..................211
    11.2.1 Chest Imaging..212
11.3 COVID-19 Classification Methodology 217
11.4 Experimental Outcomes 218
    11.4.1 Database...219
    11.4.2 Performance Metric 220
        11.4.2.1 Results: Two-Class (COVID-19 vs. Normal) Problem 221
        11.4.2.2 Results: Multi-Class (COVID-19 vs. Normal vs. Viral Pneumonia) Problem..223
    11.4.3 Results: Multi-Class Classification for Pre-Trained Network 225
11.5 Conclusion and Future Directives..228
Acknowledgment 228
References..228

11.1 INTRODUCTION

The world has been facing a global health challenge due to the novel coronavirus 2019 (COVID-19) disease, caused by a severe acute respiratory syndrome (SARS) virus (Wang, Horby, Hayden, and GaoWang et al.). Based on John Hopkins University data from 21 August 2020, there were 22,688,934 global confirmed cases including 793,773 deaths (JHU 2020). The number of confirmed cases and deaths were highest in the United States, Brazil, and India (WHO 2020). The governments of various countries imposed restrictions on movement and social distancing, and created awareness about hygiene. After June 2020, the spreading of the virus has occurred in an exponential manner.

Persons infected with the COVID-19 virus experience mild to moderate respiratory illness or pneumonia. The disease progresses through three stages: mild, severe, and critical (Assistant 2020). No symptoms, mild coughing, and fever are noticed in the mild stage. More than 50% lung involvement in imaging along with dyspnea is visualized in the severe stage. During the critical stage, failure of multiple organs and the respiratory system occurs and leads to death. Based on data from National Centre for Health Statistics (NCHS), the risk of severe illness is more than 50% in elder people (aged over 40) (CDC 2020). Persons at any age having cancer, chronic kidney problems, cardiovascular disorder, persistent respiratory disease, type-2 diabetes, and sickle cell, etc., are at increased risk for severe illness from COVID-19. A lot of clinical trials are ongoing all over the world for vaccines or potential treatments.

In order to fight against the spreading of coronavirus, effective and accurate screening and immediate medical attention to the infected person are serious requirements. To identify COVID-19-affected people, reverse transcription polymerase chain reaction (RT-PCR) is adopted all over the world. This testing method is time-consuming and suffers a high value of false-negative (FN) and false-positive (FP) rates (Ingrid et al. 2020; Shuai et al. 2020). Poor accuracy of RT-PCR may not be acceptable during the rapid spreading period of the epidemic. Many countries are facing difficulties due to an insufficient number of proper test kits. Delay in testing and inaccurate results may lead to community spread through the interaction of healthy individuals with infected ones.

Computed tomography (CT) and X-ray imaging of the chest can be utilized as an alternative to the RT-PCR test for accurate diagnosis and various stages of disease evolution (Shi et al. 2020). Easily available imaging techniques in all hospitals in India may be a quicker and cheaper method for diagnosis of COVID-19. These radiography images of affected individuals have similar lesions (Assistant 2020; Salehi et al. 2020). The most common pattern is ground glass opacity (GGO), which refers to the area of increased attenuation in the lung with preserved bronchial and vascular markings. The GGO is usually multifocal, bilateral, and peripheral. GGO may appear as a unifocal lesion (found in the anterior lobe of the right lung) during the initial stage of the disease (Salehi et al. 2020). These imaging techniques produce a large number of pathological images and need detailed analysis by radiologists. Manual evaluation of the infection is tedious, tiresome, and boring and also is influenced by individual bias and clinical experience. Sometimes, features of COVID-19 being similar to viral pneumonia may lead to false diagnosis when the healthcare system is overloaded. This leads to unnecessary utilization of healthcare resources.

Current technology makes use of artificial intelligence (AI) to effectively handle healthcare issues and complications, such as breast cancer (Shen et al. 2019), skin cancer (Keerthana and Nath 2020), and brain tumor detection (Ismael et al. 2019; Mittal et al. 2019). Deep learning (DL) techniques have triggered much interest in their application to the medical imaging domain, as they reveal detailed image features that are not possible from original images. Convolutional neural networks (CNNs) are found to be favorable in a large group of research communities for classification and detection of various pathologies from radiograph images. DL techniques are popular due to the availability of deep CNN, which achieves good performance in some applications. This technique suffers due to the lack of a large quantity of training data. Transfer learning overcomes the issue by utilizing the knowledge acquired during the training process and retraining the deep CNN with a smaller amount of data. Chouhan et al. (2020) make use of this concept for pneumonia detection in chest X-ray images using different pre-trained models such as AlexNet, DenseNet121, InceptionV3, ResNet18, and GoogleNet on ImageNet dataset. Gu et al. (2018) have used deep 19-layers VGGNet and 22-layers GoogleNet for the diagnosis of bacterial or viral pneumonia from chest radiographs. Wang et al. (2017) have used AlexNet, GoogleNet, VGG, and ResNet-50 pre-trained models on ChestX-ray8 database to diagnose eight commonly occurring thoracic pathologies: atelectasis, cardiomegaly, effusion, infiltration, mass, nodule, pneumonia, and pneumathorax. Rajpukar et al. (2018) have developed CheXNeXt with a 121-layer DenseNet architecture to identify 14 types of thoracic diseases (atelectasis, cardiomegaly, consolidation, edema, effusion, emphysema, fibrosis, hernia, infiltration, mass, nodule, pleural thickening, pneumonia, and pneumothorax) from the largest publicly available ChestX-ray14 dataset. AlexNet and GoogleNet have been used by Lakhani and Sundaram (2017) to identify pulmonary TB or normal. Yadav and Jadhav (2019) use InceptionV3 and VGG16 models for training and support vector machine (SVM) for classification on pneumonia data.

In this research work, chest images of normal, pneumonia-affected, and COVID-19-infected people have been used by a 24-layer CNN for classification of infected persons. The classification result is obtained for various optimizers and learning rates. The work is compared with the existing pre-trained networks (MobileNetV2 and SqueezeeNet, VGG16, and VGG19). The remainder of this study is organized as follows. Section 11.2 discusses COVID-19 etiology, clinical imaging features, and disease prognosis. The methodology of classification for COVID-19 is discussed in Section 11.3. Experimental outcomes are presented in Section 11.4. Finally, Section 11.5 summarizes the work and presents a roadmap for future directives.

11.2 COVID-19 ETIOLOGY, CLINICAL IMAGING FEATURES, AND PROGNOSIS

Coronavirus (CoV) is a large family of viruses, identified in the mid-1960s and named so, as the virus particles resemble a crown. It is found in six different species and seven strains (only one species divided into two strains) that cause infection in humans. These viruses are categorized into four main groups: alpha (229E and NL63), beta (OC43, HKU1, MERS-CoV, SARS-CoV), gamma, and delta (CDC 2020). Commonly people are infected by 229E, NL63, OC43, and HKU1. They are a little pathogenic and invoke mild to moderate symptoms after infection. Sometimes coronaviruses that infect animals can mutate and cause people to become sick. These are named new human coronaviruses (for example, MERSCoV, SARS-CoV, and SARS-Cov2 or COVID-19). These cause severe infections. SARS-Cov2 or COVID-19 is the seventh type of coronavirus and 85% of its genome is identical to the SARS-like virus seen in bats. These two viruses have identical amino acid sequences and protein structures. Table 11.1 summarizes the differences and similarities between SARS and COVID-19. The disease is transmitted by aerosol droplets when an infected person comes into contact with a healthy individual. Exposed

TABLE 11.1
Silent Points between SARS vs. COVID-19

Description or Item	SARS	COVID-19
Virus	Coronavirus	Coronavirus
Origin	Guangdong, China, November 2020	Wuhan, China, December 2020
Natural reservoir	Bat	Bat
Intermediate host	Civet cats	Pangolins
Predominant cellular receptor	ACE2	ACE2
Incubation period (days)	2–10	1–14, Mostly 3–7
Reproduction number (R0)	1.4–2.5 (median 1.95)	3
Symptoms	Fever, head and body aches, cough, shortness of breath, diarrhea, sore throat	Most common: fever, dry cough, tiredness; less common: aches and pains, sore throat, diarrhea, pink eye, rash on skin or discoloration of fingers or toes, headache, nausea, loss of taste or smell; serious: difficult breathing, chest pain, loss of speech
Fatality rate	10%	Evolving
Risky	People above 60 or having health issues	Elderly persons, people with immunodeficiency or chronic disease

Sources: CDC 2020, Caldaria et al. 2020, Cascella et al. 2020.
Abbreviations: SARS: Severe Acute Respiratory Syndrome; ACE: Angiotensin-converting enzyme.

mucus from the infected person (through coughing and sneezing), naked eye, exposed to contaminated aerosols in a closed environment, increases the chances of infection (WHO 2020). The disease shows symptoms of pneumonia, which affects the lungs. So, analysis of chest images can be a potential tool to diagnose the disease. Chest imaging includes X-ray, CT scan, and lung ultrasound. Details about the features in chest image during infection in various modalities are discussed in the following subsection.

11.2.1 CHEST IMAGING

X-ray images of the chest hardly show any changes at the initial stage of the infection and result in false-negative results. At the advanced stage of the disease, bilateral multifocal alveolar opacities are noticed. This in later stages leads to junction of complete opacity of the lung (Cascella et al. 2020; Chen et al. 2020). Excess fluid between the layers of pleura outside the lungs (pleural effusion) can be visualized. Figure 11.1 shows the chest images of a 72-year-old lady, who had been admitted to hospital due to acute respiratory failure and symptoms of COVID-19 (Radiology Assistant 2020).

Diagnosis of Coronavirus from Radiograph Images

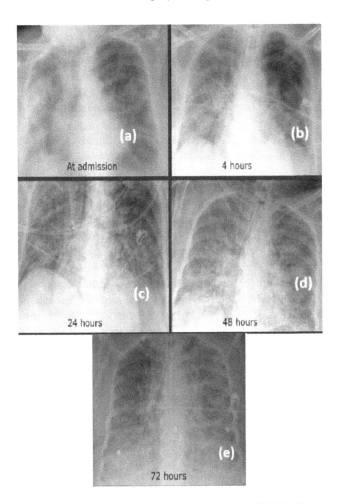

FIGURE 11.1 X-ray image features at various stages of COVID-19 infected person: (a) image at admission, (b) image after 4 hours, (c) image after 24 hours, (d) image after 48 hours, and (e) image after 72 hours. Images are taken from Radiology Assistant: COVID-19 Imaging Findings (2020).

Figure 11.1a shows bilateral alveolar consolidation with peripheral distribution and it worsened after 4 hours (shown in Figure 11.1b). Over 24–48 hours, bilateral alveolar consolidation becomes more prominent. After 72 hours of the infection, the chest image shows acute respiratory distress syndrome (ARDS). It is shown in Figure 11.1e. At this stage, the patient died due to multi-organ failure in the ICU.

CT images can be mainly used for infected COVID-19 cases due to their higher sensitivity and resolution than X-ray images. Commonly found patterns/features are ground glass opacity (GGO), peripheral or posterior distribution, crazy paving, air bronchi, fibrous lesion, vascular thickening, and halo sign (Assistant 2020; Chen et al. 2020).

GGO is the commonly found denser and more blurred pattern in lung images for a COVID-19 patient. Blood vessels are visible here. In the early stage, GGO appears as unifocal and is located in the inferior lobe of the right lung. Here, pulmonary blood vessels are visible. In the later stage, it appears as multifocal, bilateral, and peripheral. Pathological change in GGO represents invading of the virus over bronchioles and alveolar epithelium. It starts replicating and causes the alveolar cavity to leak. This thickens the alveolar wall of the lung.

White lung is visible due to the participation of the alveoli and mucosal ulcers during the inflammation process. As the disease progresses, air bronchus signs (i.e., low-density shadowing) are visible. This leads to thickening and swelling of the bronchial wall without blocking of the bronchioles.

Paving is visualized at the later stage of infection and has an identical pattern to paving stones. It appears due to the superimposition of thickened lobular interval, interlobular line, and GGO. *Vascular thickening* can be visualized at all stages of the infection. *Fibrous lesions* appear hyperplasia of the lung and replace the normal cellular components to form scars. Density of lesions decreases from the center to the edges and creates a circle of cloud in GGO called a *halo*. It has non-uniform thickness. Figure 11.2 shows the various image features seen in CT scans for a COVID-19 patient at different stages. Arrows mark the features. During disease progression, pleural effusion, lymphadenopathy, cavitation, CT halo sign, and pneumothorax are visible in CT images.

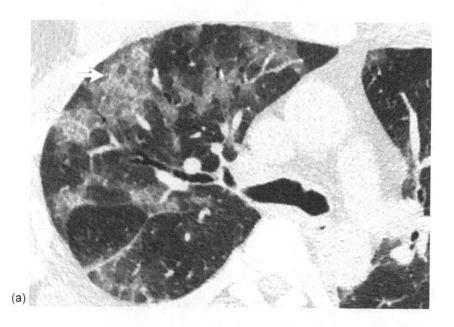

(a)

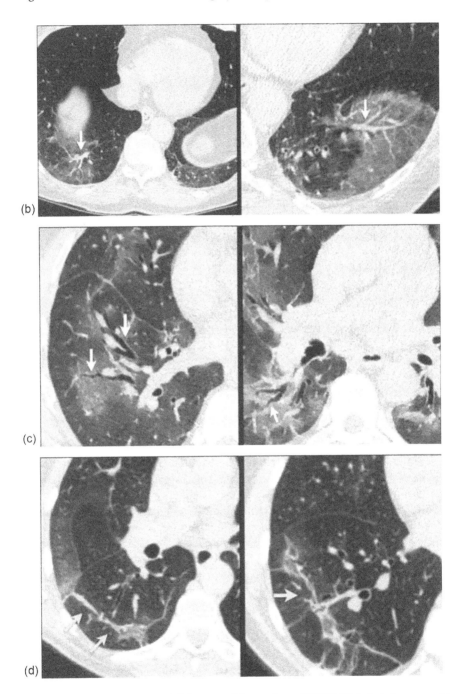

FIGURE 11.2 Features from CT images of lungs from COVID-19 infected person: (a) paving, (b) vascular dilation, (c) bronchiectasis due to fibrous lesions, and (d) architectural distortion due to subpleural bands. Images are taken from Radiology Assistant: COVID-19 Imaging Findings (2020).

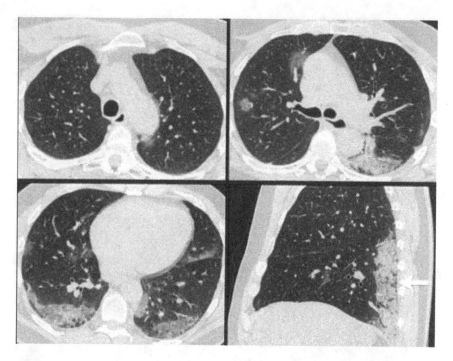

FIGURE 11.3 CT imaging of lungs for susceptible a COVID-19 patient, whose RT-PCR was negative. Image has been taken from Radiology Assistant: COVID-19 Imaging Findings (2020).

Figure 11.3 shows the lung CT image of a 59-year-old male, whose RT-PCR test was found to be negative. This image was captured from the patient due to complaints of fever and non-productive cough. This image shows GGO in the posterior part (represented by the pale arrow). This patient's PCR becomes positive for COVID-19 after 2 days. By comparing, the CT may be an efficient modality for COVID-19 detection. GGO is visible in CT images (represented by the dark arrow) and not in X-ray (Figure 11.4).

The lung ultrasound approach can be used during the first 24 hours from suspecting the disease (Cascella et al. 2020). This imaging modality will help to study the progression of infection in newborns, children, and adults (Xing et al. 2020). The main features are pleural lines, B-lines, thickening, deep vein occlusion, and lung consolidation. Pleural lines are thick, discontinuous, and irregular. B-lines can cascade to form white lungs. In the early stage of the disease, B-lines are visible. As the disease progresses, white lung and thickening occur which further lead to consolidations. Proportionate changes in B-lines and consolidation can be seen with regard to the severity of the disease. B-lines and confluent B-lines are prominent features in severe conditions. Deep vein occlusion and pleural ebullition are significant signs in ultrasound images for critical COVID-19 patients. Suspect cases show the light beam in lung ultrasound (Volpicelli and Gargani 2020).

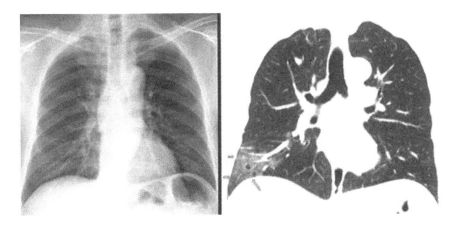

FIGURE 11.4 Comparison of X-ray (left) and CT (right) image of an infected person. Image has been taken from Radiology Assistant: COVID-19 Imaging Findings (2020).

11.3 COVID-19 CLASSIFICATION METHODOLOGY

In the first half of 2020, several research papers have been reported for diagnosis of COVID-19 from the CT and X-ray images using DL techniques. Few of them have reported their result on a very small dataset (images less than a thousand), which may not be retained on a large dataset. Many researchers have used transfer learning for the detection of the disease.

Togacar et al. (2020) have used the strength of MobileNetV2 and SqueezeNet for feature extraction and support vector machine (SVM) classifier for classification of COVID-19 chest X-ray images. They have used fuzzy color technique as a pre-processing stage for removal of noise and stacking technique for combining the original data with fuzzy color images. The stacking dataset was trained by existing DL models (MobileNetV2 and Squeezenet) for feature extraction. From these, efficient features have been extracted by social mimic optimization for classification by SVM. The authors have used three-class datasets (COVID-19, normal, and pneumonia) obtained from Qatar University (X-ray images) and Montreal University (CT images). The authors have reported an accuracy of 99.27% for 1,357 features and fivefold validation. Makris et al. (2020) have used different pre-trained networks (VGG16, VGG19, MobileNetV2, InceptionV3, Xception, InceptionResnetV2, DenseNet201, Resnet152V2, and NASNetLarge) on X-ray images obtained from Kaggle's repository. They found the highest overall accuracy of 95.88% for VGG16 network and the lowest overall accuracy of 38.23% for DenseNet201. MobileNetV2, InceptionV3, Xception, InceptionResnetV2, DenseNet201, and Resnet152V2 have less than 80% accuracy but 100% precision.

Fan et al. (2020) have used Inf-Net for identification of coronavirus infection from the CT image slices. The authors aggregated the features from a parallel partial decoder for global mapping. This network reverses attention and explicit edge-attention for accurate identification of the infected regions. They reported the specificity of 97.4% and 97.7% for Inf-Net and Semi-Inf-Net network. This method clearly identifies the objects in poor contrast regions among the infected and non-infected lung tissues. This method may provide better performance by dropping the non-infected slices. Singh et al. (2020) have used CT image slices and multi-objective differential evolution (MODE)-based CNN for COVID-19 classification. The authors have compared their method with adaptive neuro-fuzzy inference systems (ANFIS) and artificial neural networks (ANN) and found an improvement of nearly 2% in F-score and sensitivity.

Ozturk et al. (2020) have proposed a 19-layer (17 convolution layers) CNN architecture for classification of COVID-19 from the chest X-ray images (Cohen et al. 2020). This network uses fewer layers of Darknet19 followed by batch normalization (helps in standardizing the input, reducing training time, and increases stability). Leaky ReLU prevents the death of the neuron. This network reported an accuracy of 98.08%.

Elaziz et al. (2020) have proposed a classification method which depends on orthogonal moment features and their selection by Manta-Ray Foraging Optimization (MRFO) based on differential evolution (DE). The authors evaluated this method on X-ray images from Joseph Paul Cohen repository and Qatar University and achieved an accuracy of 98.91%.

In this chapter, a 24-layer CNN architecture consists of six convolution layers is proposed for classification of chest images infected by COVID-19. All the input images are resized to a single channel with 256×256 pixels before being fed into the network for making the training faster. Two filter sizes, 3×3 and 5×5, have been used for convolution. ReLU activation function is used for non-linearity, after the batch normalization (BN). ReLU does not saturate and the gradient remains high always. BN standardizes the input to a layer for each mini-batch and accelerate the training process. It reduces the number of epochs by stabilizing the learning process. Max pooling uses a filter size of 2×2 and shortens the computation time by cutting down the spatial dimension. It does not contribute to learning. The result of the last pooling process is fed to the fully connected layer for final classification. Then, softmax is used just before the output layer, which assigns probabilities to each class. The general structure of the proposed CNN network is represented in Table 11.2.

11.4 EXPERIMENTAL OUTCOMES

In this study, two different schemes are studied for the classification of COVID-19, pneumonia, and normal images, using different pre-trained models and the proposed 24-layer CNN network. Various publicly available datasets are used for this study. The descriptions of various datasets are given in Subsection 11.4.1. These datasets have a smaller number of images, which may affect the proposed network accuracy for a large dataset. So, data augmentation is carried out to increase the number of images before being fed to the network. Training was performed with and without augmentation for different optimizers, learning rate (LR), and epochs. For various hyper-parameters, different performance metrics are computed and discussed in Subsection 11.4.2.

TABLE 11.2
General Structure and Parameters of Proposed Architecture

Type	Stride	Filter Size	Output Size
Input	–	–	256×256×1
Convolution1	1	3×3×1×16	254×254×16
BN1	–	–	254×254×16
ReLu1	–	–	254×254×16
Max pooling1	2	–	127×127×16
Convolution2	1	3×3×16×32	125×125×32
BN2	–	–	125×125×32
ReLu2	–	–	125×125×32
Max Pooling2	–	–	62×62×32
Convolution3	1	3×3×32×64	60×60×64
BN3	–	–	60×60×64
Convolution4	1	3×3×64×128	58×58×128
BN4	–	–	58×58×128
ReLu4	–	–	58×58×128
Max pooling4	2	–	29×29×128
Convolution5	1	5×5×128×128	25×25×128
BN5	–	–	25×25×128
Convolution6	1	5×5×128×128	21×21×128
BN6	–	–	21×21×128
ReLu6	2	–	21×21×128
Max pooling6	2	–	10×10×128
Fully Connected	–	3×128000	1×1×3
Soft Max	–	–	1×1×3

11.4.1 DATABASE

X-ray and computed tomography (CT) images of the human lung help the physicians for a quick diagnosis of the disease. A couple of CT and X-ray image databases are publicly available. Joseph Paul Cohen from the University of Montreal first shared a COVID-19 database on the GitHub website (Cohen et al. 2020). It contains a total of number 910 X-rays and CT scans of COVID-19 along with MERS, SARS, and ARDS. The images are available in jpg file format with different dimensions. It has 76 posteroanterior (PA), 11 anteroposterior (AP), and 13 AP Supine (laying down) severe acute respiratory syndrome-related coronavirus 2 (SARSr-CoV-2) or COVID-19 images.

Researchers from Qatar University Doha, Qatar, and the University of Dhaka, Bangladesh, in association with the medical practitioners from Pakistan and Malaysia, created a database of chest X-ray images for COVID-19 affected persons along with normal and pneumonia-affected people (T. Rahman and Chowdhury 2020). It contains 219 COVID-19, 1,341 normal, and 1,345 viral pneumonia chest

X-ray images, having the dimension of 1024×1024 in portable network graphics (.png) file format. It has three classes and is more suitable for the study of COVID-19.

Italian Society of Medical and Interventional Radiology (ISMIR) has developed a COVID-19 database from 68 patients, which consists of 384 lung images (94 X-ray and 290 CT) with different resolutions (www.sirm.org/en/2020/) (ISMIR 2020).

Jinyu Zhao et al. have created a COVID-CT-dataset, containing 349 COVID and 397 non-COVID CT images from 216 patients (Zhao et al. 2020). Among different imaging modalities, CT images of the chest provide a three-dimensional picture of the chest lung from the slices. This helps to detect the sign of infection in the early stage of the disease and its nature in the later stage. Analysis of the slices from CT images provides qualitative evaluation to fight against COVID-19. Some of the CT slices features related to the infections are not visible.

11.4.2 Performance Metric

The performance of CNN is measured by accuracy, precision, sensitivity, specificity, and dice similarity coefficient (DSC). For any class, they can be defined in Equation 11.1. The classes are COVID-19 and normal for a two-class problem; COVID-19, normal, and viral pneumonia for multi-class. True positive (TP) and true negative (TN) represent correctly classified COVID-19 and non-COVID-19 images, respectively. Wrongly detected samples are represented as false negative (FN) and false positive (FP), respectively. Accuracy represents the discriminating capability of the classifier. Precision or positive predicative value (PPV) measures the correctly predicted labels to the total number of target labels. Whereas, recall (sensitivity) measures the correctly predicted labels to the total number of predicated labels. F-score represents the harmonic mean between precision and recall. The value of DSC is the same as F1-score. The performance values range from 0 to 1. The metric value of these measures should lie close to "1" for better classification.

$$\text{Accuracy} = \frac{TP + TN}{TP + TN + FP + FN}$$

$$\text{Precision} = \frac{TP}{TP + FP} = PPV$$

$$\text{Sensitivity} = \frac{TP}{TP + FN} = \text{Recall}$$

$$\text{Specificity} = \frac{TN}{TN + FP}$$

$$DSC = \frac{2*TP}{2*TP + FP + FN}$$

$$F1-score = 2*\frac{\text{Precision}*\text{Sensitivity}}{\text{Precision} + \text{Sensitivity}}$$

(11.1)

Micro-F1 and Macro-F1 are computed for more than two classes. Micro-F1 is the same as the overall accuracy or total precision or total recall of the model. Macro-F1 is evaluated as per Equation 11.2, where C denotes the number of classes, P_i is the precision of each ith class, n_i denotes number of samples in the ith class, and N denotes the total number of samples.

$$\text{Macro} - F1 = \frac{\sum_{i=1}^{C} P_i}{C}$$
$$\text{Macro} - F1 = \frac{\sum_{i=1}^{C} P_i \times n_i}{N}$$
(11.2)

The performance metric is computed for two-class (COVID-19 vs. normal) and multi-class (COVID-19 vs. normal vs. viral pneumonia) classification. The simulation is carried out using MATLAB 2020, a software with a deep learning toolbox. The hardware specification is: Intel core i7-7700 64-bit processor (3.6 GHz), with 8GB RAM and 2GB NVDIA GeForce GT1030 graphics card.

11.4.2.1 Results: Two-Class (COVID-19 vs. Normal) Problem

Performance of the proposed CNN architecture for two-class (COVID-19 vs. normal) classification is given in Table 11.3. The result is obtained for X-ray images taken from Kaggle database (T. Rahman 2020). Out of 219 COVID and 1,341 normal images, 80% from both the classes are used for training. Rest 20% of the images are used for testing. The experiments are carried out for different optimizers (Sgdm, Adam, and RmsProp), epochs (5, 10, and 15), and learning rates (0.1, 0.01, and 0.001). All the optimizers recorded higher accuracy for the learning rate 0.001. This value was found to be 99.68%, 99.68%, and 99.40% for Sgdm, Adam, and RmsProp optimizers, respectively. In all the optimizers, accuracy value increased with lower learning rate and higher epoch. A decline train is seen if the epoch is increased beyond a certain value. This may occur due to overfitting. A highest F1-score of 98.88% is observed for Adam optimizer with 5 epochs and 0.001 learning rate. To check the perfection in the classification of the proposed network, augmented data is used. Here, the number of images in the dataset is increased by rotation operation for all the classes before feeding into the network. A total of 19,980 augmented images (6570 COVID-19 and 13,410 normal) are fed to the network for classification. Accuracy of 99.64% is obtained for the Adam optimizer with a learning rate of 0.001.

The proposed 24-layer CNN network is used for CT images taken from (Zhao et al. 2020). For a 0.001 learning rate, a highest accuracy of 71.81% is obtained by the Sgdm optimizer. The proposed network attains an accuracy of 61.07% and 56.4% for Adam and RmsProp optimizers, respectively. The classification performance in CT images is tabulated in Table 11.4. Accuracy is recorded to be less. This may be due to the non-availability of the pathological features in some of the slices, which are present in the COVID-19 class. It may be due to the initial stage of the disease and features may not be prominent in any of the slices in CT images.

TABLE 11.3
Two-Class Classification Performance Metric for X-Ray Chest Images: Proposed 24-Layer CNN for Different Optimizers (80% Training and 20% Testing)

Optimizer	Learning Rate	Epoch	Accuracy	Precision	Sensitivity	Specificity	DSC	F1-Score
Sgdm	0.001	5	98.72	95.45	95.45	99.25	95.45	95.45
		10	**99.68**	97.78	100	99.63	**98.88**	98.88
		15	98.39	100	88.64	100	93.98	93.98
Adam	0.001	5	**99.68**	97.78	100	99.63	98.88	**98.88**
		10	99.36	100	95.45	100	97.67	97.67
		15	99.36	95.65	100	99.25	97.78	97.78
RmsProp	0.001	5	**99.36**	100	95.45	100	97.67	**97.67**
		10	97.44	90.91	90.91	98.51	90.91	90.91
		15	99.04	97.67	95.45	99.63	96.55	96.55

TABLE 11.4
Two-Class Classification Performance Metric for CT Images: Proposed 24-Layer CNN for Different Optimizers (80% Training and 20% Testing)

Optimizer	Learning Rate	Epoch	Accuracy	Precision	Sensitivity	Specificity	DSC	F1-Score
Sgdm	0.001	10	**71.81**	68.92	72.86	70.89	70.83	70.83
Adam	0.001	5	61.07	65.00	37.14	82.28	47.27	47.27
RmsProp	0.001	5	56.4	51.91	97.14	20.25	67.66	67.66

TABLE 11.5
Multi-Class Classification Performance Metric for Infected X-Ray Chest Images: Proposed 24-Layer CNN for Different Optimizer (80% Training and 20% Testing)

| Optimizer | LR | Epoch | Class Precision | | | Class Recall | | | Accuracy | Macro-F1 |
			C1	C2	C3	C1	C2	C3		
Sgdm	0.00001	25	76.3	93.6	89.5	65.9	93.3	91.8	90.5	86.47
	0.0001	25	97.6	94.4	94.1	93.2	95.1	94.1	94.5	95.37
	0.001	25	93.5	95.9	97.4	97.7	96.6	95.9	**96.4**	**95.6**
	0.01	25	95.6	94.8	95.1	97.7	95.5	94.1	95.0	95.17
	0.1	25	NaN	0	46.4	0	0	100	46.3	NaN
RmsProp	0.00001	25	94.9	96.6	93.2	84.1	95.1	96.3	94.8	94.9
	0.0001	25	94.9	94.6	95.5	84.1	98.1	93.7	95	95
	0.001	25	97.4	96.9	91.9	86.4	92.9	97.4	94.5	95.4
	0.01	25	86.4	96.3	95.5	86.4	96.3	95.5	**95.2**	92.73
	0.1	25	100	100	48.6	2.3	9.7	100	50.9	82.86
Adam	0.00001	25	90.9	96.6	94.2	90.9	94.8	95.9	95.0	93.9
	0.0001	25	90.9	96.0	96.2	90.9	97.4	94.8	**95.7**	94.37
	0.001	25	93.0	95.9	94.5	90.9	95.1	95.5	95.5	94.47
	0.01	25	83.3	95.3	97.3	90.9	97.4	93.7	95.2	91.97
	0.1	25	90.9	89.8	96.3	90.9	98.1	87.4	92.6	92.33

Notations: C1-COVID-19, C2-Non-COVID, C3-Viral Pneumonia.

11.4.2.2 Results: Multi-Class (COVID-19 vs. Normal vs. Viral Pneumonia) Problem

Table 11.5 shows the various performance values for three classes (i.e., COVID-19, normal, and viral pneumonia). Eighty percent and 20% of the images are considered for training and testing, respectively. Highest accuracies of 96.4%, 95.2%, and 95.7% are recorded for Sgdm, RmsProp, and Adam optimizers, respectively. A highest value of Macro-F1 is noticed in the Sgdm optimizer for a learning rate of 0.001. Classification performance is found to be less in other hyper-parameters. Figure 11.5 shows the confusion matrix for multi-class classification from the X-ray images for Sgdm, RmsProp, and Adam optimizers. For 0.001 learning rate and 25 epochs, the highest accuracy is obtained for the Sgdm optimizer.

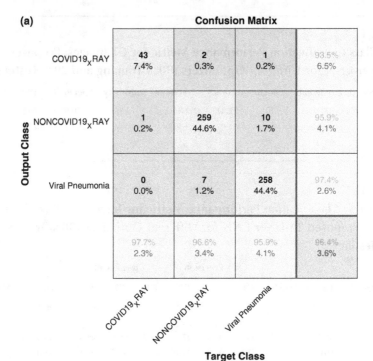

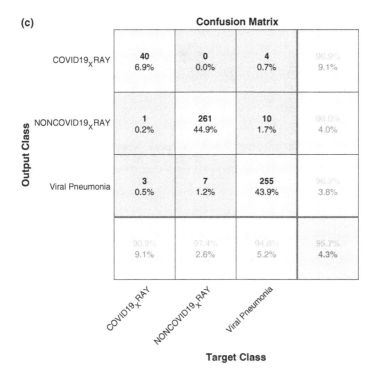

FIGURE 11.5 Confusion matrix for multi-class classification of X-ray images by the proposed CNN having validation frequency = 3 for various optimizers: (a) Sgdm (LR = 0.001), (b) RmsProp (LR = 0.01), and (c) Adam (LR = 0.0001).

11.4.3 Results: Multi-Class Classification for Pre-Trained Network

Makris et al. (2020) have used different pre-trained networks (VGG16, VGG19, MobileNetV2, InceptionV3, Xception, InceptionResNetV2, DenseNet201, ResNet152V2, and NASNetLarge) for the classification of COVID-19 from X-ray images of chest. The authors found that VGG16 and VGG19 provide the highest accuracy compared to other networks. MobileNetV2 and DenseNet201 have very low performance in terms of accuracy. But these two networks have the highest precision value (100%) for the COVID-19 class. Here, two different pre-trained networks (SqueezeNet and ResNet18) have been used along with VGG16 and VGG19 networks for the study. These four networks have been used for classification with different hyper-parameters. Performance metric values for three-class (i.e., COVID-19 vs. normal vs. viral pneumonia) classification are represented in Table 11.6. The experiment is conducted with 25 epochs for Sgdm optimizers. The performance is computed for different learning rates (LR). SequeezeNet and ResNet18 provide an overall accuracy of 96.9% and 97.9%, respectively, for a 0.001 learning rate. In contrast, for VGG16 and VGG19 the accuracy is comparably less. Classification performances of the pre-trained networks (SequeezeNet and ResNet18) are shown by confusion matrix, in Figure 11.6. Results are obtained for 25 epochs and 0.001 learning rate and show 100% precision.

TABLE 11.6
Multi-class Classification Performance Metric for Infected X-Ray Chest Images by Various Pre-Trained Network (80% Training and 20% Testing)

Network	LR	Class Precision			Class Recall			Accuracy	Macro-F1
		C1	C2	C3	C1	C2	C3		
SqueezeNet	0.00001	90.9	90.5	92.0	68.2	95.9	90.3	91.2	91.13
	0.0001	97.6	94.9	97.3	90.9	97.8	95.5	96.2	96.6
	0.001	**100**	95.0	98.5	93.2	98.9	95.5	**96.9**	97.8
	0.01	7.6	NaN	NaN	100	0	0	7.6	NaN
ResNet18	0.00001	92.6	88.0	94.9	56.8	98.5	89.6	91.2	91.83
	0.0001	100	92.3	98.0	97.7	98.1	92.2	95.4	96.76
	0.001	**100**	96.4	99.2	100	99.3	96.3	**97.9**	98.53
	0.01	100	93.6	98.8	100	98.9	93.3	96.4	97.47
	0.1	78.8	95.6	96.9	93.2	97.4	92.2	94.7	90.43

Notations: C1-COVID-19, C2-Non-COVID, C3-Viral Pneumonia.

Diagnosis of Coronavirus from Radiograph Images

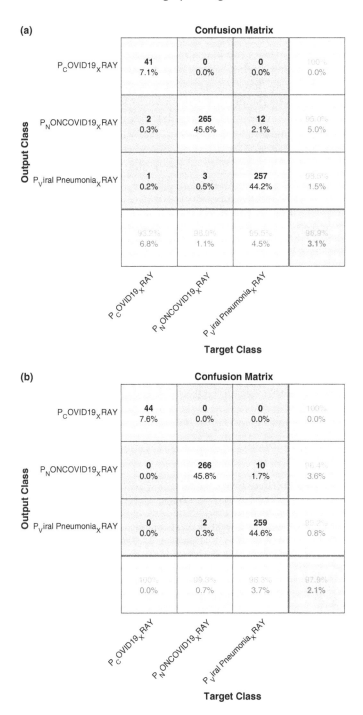

FIGURE 11.6 Confusion matrix for multi-class classification of X-ray images by pretrained networks having validation frequency = 3, LR = 0.001: (a) SquezeezeNet and (b) ResNet18.

11.5 CONCLUSION AND FUTURE DIRECTIVES

COVID-19 infection affects the lungs and changes can be visualized from lung images. In this chapter, details about the features of COVID-19 along with some detection methods have been discussed. AI-based COVID-19 classification has been studied by the pre-trained networks and proposed a 24-layer CNN architecture. For this study, publicly available X-ray and CT images are used. Experiments were conducted for two-class (COVID-19 vs. normal) and multi-class (COVID-19 vs. normal vs. viral pneumonia) for various epochs, learning rates, and optimizers. Proposed CNN network attains an accuracy of 99.68% and 71.81% for X-ray and CT database, respectively, for two classes. Performance of classification in CT images can be improved with preprocessing, which may help to capture the feature in infected slices of CT images. For multi-class, an accuracy of 96.4% is recorded for the Sgdm optimizer with a learning rate of 0.001. The accuracy is found to be less for other learning rates and optimizers. Different pre-trained networks have been used for classification. SquezzeNet and ResNet18 have achieved an overall accuracy of 96.9% and 97.9%, respectively. Both these networks recorded 100% precision. Classification accuracy of X-ray images in proposed CNN architecture is better compared to the methods mentioned in the literature.

ACKNOWLEDGMENT

This work has been carried at National Institute of Technology Puducherry, Karaikal, India.

There is no conflict of interest.

REFERENCES

Radiology Assistant: COVID-19 Imaging Findings. https://radiologyassistant.nl/chest/covid-19/covid19-imaging-findings (accessed August 21, 2020).

CDC People with Certain Medical Conditions. https://www.cdc.gov/coronavirus/2019-ncov/need-extra-precautions/people-withmedical-conditions.html (accessed August 21, 2020).

Caldaria, A., C. Conforti, N. D. Meo, C. Dianzani, M. Jafferany, T. Lotti, I. Zalaudek, and R. Giuffrida (2020). COVID-19 and SARS: differences and similarities. *Dermatologic Therapy*. https://doi.org/10.1111/dth.13395.

Cascella, M., M. Rajnik, A. Cuomo, S. C. Dulebohn, and R. D. Napoli (2020). Features, evaluation, and treatment of coronavirus (COVID-19). *StatPearls* [Internet]. https://www.ncbi.nlm.nih.gov/books/NBK554776/.

Chen, H., L. Ai, H. Lu, and H. Li (2020). Clinical and imaging features of COVID-19. *Radiology of Infectious Diseases*. https://doi.org/10.1016/j.jrid.2020.04.003.

Chouhan, V., S. K. Singh, A. Khamparia, D. Gupta, P. Tiwari, C. Moreira, and V. H. C. D. A. Robertas Damaševičius (2020). A novel transfer learning-based approach for pneumonia detection in chest X-ray images. *Applied Sciences*. https://doi.org/10.3390/app10020559.

Cohen, J. P., P. Morrison, and L. Dao. COVID-19 image data collection. https://github.com/ieee8023/covid-chestxray-dataset (accessed August 20, 2020).

Cohen, J. P., P. Morrison, L. Dao, K. Roth, T. Q. Duong, and M. Ghassemi. COVID-19 image data collection: prospective predictions are the future. https://github.com/ieee8023/covid-chestxray-dataset (accessed August 20, 2020).

Elaziz, M. A., K. M. Hosny, A. Salah, M. M. D. S. Lu, and A. T. Sahlol (2020, June). New machine learning method for image based diagnosis of COVID-19. *PLOS ONE* 15(6). https://doi.org/10.1371/journal.pone.0235187.

Fan, D.-P., G.-P. J. Tao Zhou, G. C. Yi Zhou, H. Fu, J. Shen, and L. Shao (2020, May). InfNet: Automatic COVID-19 Lung Infection segmentation from CT images. https://arxiv.org/pdf/2004.14133.pdf.

Gu, X., L. Pan, H. Liang, and R. Yang (2018, June). Classification of bacterial and viral childhood pneumonia using deep learning in chest radiography. *Proceedings of the 3rd International Conference on Multimedia and Image Processing.* Beijing Normal University, Zhuhai, China, pp. 88–93. https://doi.org/10.1145/3195588.3195597.

Ingrid Arevalo-Rodriguez, Diana Buitrago-Garcia, Daniel Simancas-Racines, Paula Zambrano-Achig, Rosa Del Campo, Agustin Ciapponi, Omar Sued, Laura Martinez-García, Anne W. Rutjes, Nicola Low, Patrick M. Bossuyt, Jose A. Perez-Molina, Javier Zamora (2020). False-negative results of initial RT-PCR assays for COVID-19: a systematic review. *medRxiv.*

Ismael, S. A. A., A. Mohammed, and H. Hefny (2019). An enhanced deep learning approach for brain cancer MRI images classification using residual networks. *Artificial Intelligence in Medicine.* https://doi.org/10.1016/j.artmed.2019.101779.

ISMIR (Italian Society of Medical and Interventional Radiology). *Archivi Annuali: 2020.* https://www.sirm.org/en/2020/ (accessed August 26, 2020).

JHU COVID-19 Dashboard by the Center for System Science Engineering (CSSE) at John Hopkins University. https://coronavirus.jhu.edu/map.html (accessed August 21, 2020).

Keerthana, D., and M. K. Nath (2020, June). A technical review report on deep learning approach for skin cancer detection and segmentation. In Springer (Ed.), *Springer Lecture Notes on Data Engineering and Communications Technologies.* Jan Wyzykowski University, Poland and B.M. Institute of Engineering and Technology, India. International Conference on Data Analytics and Management (ICDAM-2020).

Lakhani, P., and B. Sundaram (2017). Deep learning at chest radiography: automated classification of pulmonary tuberculosis by using convolutional neural networks. *Radiology.* https://doi.org/10.1148/radiol.2017162326.

Makris, A., I. Kontopoulos, and K. Tserpes (2020, May). Covid-19 detection from chest X-ray images using deep learning and convolutional neural networks. https://doi.org/10.1101/2020.05.22.20110817.

Mittal, M., L. M. Goyal, S. Kaur, I. Kaur, A. Verma, and D. J. Hemanth (2019). Deep learning based enhanced tumor segmentation approach for MR brain images. *Applied Soft Computing Journal.* https://doi.org/10.1016/j.asoc.2019.02.036.

Ozturk, T., M. Talo, A. Yildirim, U. B. Baloglu, O. Yildirim, and U. R. Achrya (2020, April). Automated detection of COVID-19 cases using deep neural network with X-ray images. *Computers in Biology and Medicine* 121(103792). https://doi.org/10.1016/j.compbiomed.2020.103792.

Rahman, T., and M. Chowdhury. A. K. COVID-19 radiography database, Kaggle. https://www.kaggle.com/tawsifurrahman/covid19-radiography-database/data (accessed August 20, 2020).

Rajpurkar, P., J. Irvin, R. L. Ball, K. Zhu, B. Yang, H. Mehta, T. Duan, D. Ding, A. Bagul, C. P. Langlotz, B. N. Patel, K. W. Yeom, K. Shpanskaya, F. G. Blankenberg, J. Seekins, T. J. Amrhein, D. A. Mong, S. S. Halabi, E. J. Zucker, A. Y. Ng, and M. P. Lungren (2018). Deep learning for chest radiograph diagnosis: a retrospective comparison of the chexnext algorithm to practicing radiologists. *PLOS Medicine.* https://doi.org/10.1371/journal.pmed.1002686.

Salehi, S., A. Abedi, S. Balakrishnan, and A. Gholamrezanezhad (2020, July). Coronavirus disease 2019 (COVID-19): a systematic review of imaging findings in 919 patients. *AJR. American Journal of Roentgenology* 215, 87–93. https://www.ajronline.org/doi/full/10.2214/AJR.20.23034.

Shen, L., L. R. Margolies, J. H. Rothstein, E. Fluder, R. McBride, and W. Sieh (2019). Deep learning to improve breast cancer detection on screening mammography. *Scientific Reports.* https://doi.org/10.1038/s41598-019-48995-4.

Shi, F., J. Wang, J. Shi, Z. Wu, Q. Wang, Z. Tang, K. He, Y. Shi, and D. Shen (2020). Review of artificial intelligence techniques in imaging data acquisition, segmentation and diagnosis for COVID-19. *IEEE Reviews in Biomedical Engineering,* 14, 1.

Shuai, W., K. Bo, M. Jinlu, Z. Xianjun, X. Mingming, G. Jia, C. Mengjiao, Y. Jingyi, L. Yaodong, M. Xiangfei, and X. Bo (2020). A deep learning algorithm using CT images to screen for corona virus disease (COVID-19). medRxiv. https://www.medrxiv.org/content/early/2020/04/24/2020.02.14.20023028/.

Singh, D., V. Kumar, Vaishali, and M. Kaur (2020, April). Classification of COVID-19 patients from chest CT images using multi-objective differential evolution–based convolutional neural networks. *European Journal of Clinical Microbiology and Infectious Diseases* 39, 1379–1389. https://doi.org/10.1007/s10096-020-03901-z.

Togacar, M., B. Ergen, and Z. Comert (2020, May). COVID-19 detection using deep learning models to exploit social mimic optimization and structured chest x-ray images using fuzzy color and stacking approaches. *Computers in Biology and Medicine* 121. https://doi.org/10.1016/j.compbiomed.2020.103805.

Volpicelli, G., and L. Gargani (2020). Sonographic signs and patterns of COVID-19 pneumonia. *Ultrasound.* https://doi.org/10.1186/s13089-020-00171-w.

Wang, C., P. W. Horby, F. G. Hayden, and G. F. Gao. A novel coronavirus outbreak of global health concern. *Lancet* 395(10223), 470–473. https://doi.org/10.1016/S0140-6736(20)30185-9.

Wang, X., Y. Peng, L. Lu, Z. Lu, M. Bagheri, and R. M. Summers (2017, July). Chestx-ray8: hospitalscale chest x-ray database and benchmarks on weakly-supervised classification and localization of common thorax diseases. In *IEEE Conference on Computer Vision and Pattern Recognition (CVPR),* Honolulu, HI, IEEE, pp. 2097–2106. https://ieeexplore.ieee.org/document/8099852.

WHO. WHO coronavirus disease (COVID-19) dashboard. https://covid19.who.int/ (accessed August 21, 2020).

Xing, C., Q. Li, H. Du, W. Kang, J. Lian, and L. Yuan (2020). Lung ultrasound findings in patients with COVID-19 pneumonia. *Critical Care.* https://doi.org/10.1186/s13054-020-02876-9.

Yadav, S. S. and S. M. Jadhav (2019). Deep convolutional neural network based medical image classification for disease diagnosis. *Journal of Big Data.* https://doi.org/10.1186/s40537-019-0276-2.

Zhao, J., Y. Zhang, X. He, and P. Xie. COVID-CT-dataset: a CT scan dataset about COVID-19. https://github.com/UCSD-AI4H/COVID-CT (accessed August 20, 2020).

12 Data Analytics for COVID-19

Shreyas Mishra

CONTENTS

12.1 Introduction 231
12.2 Literature Study 232
12.3 Exploratory Data Analysis.. 233
12.4 Compartmental Models for Epidemiology 235
12.5 Deep Learning for COVID-19 Diagnosis... 238
12.6 Survival Analysis for COVID-19.. 243
12.7 COVID-19 Forecasting Models 245
12.8 Conclusion 248
References.. 251

12.1 INTRODUCTION

The coronavirus is a family of viruses that are named after their spiky crown. In September 2019, about 1 million cases were emerging every five days. In India, till September 30, 2020, about 6.3 million people were infected, with about 80,000 cases and 1,000 deaths being added per day, although the number of cases per capita was far lower than most countries. This chapter will provide some basic global epidemiological and trend analysis of the virus. Data analysis on COVID-19 is necessary to track the spread of the virus and its impact on communities as well as to forecast future predictions and suggest healthcare measures and responses. The objective of this chapter is to provide an analysis of various parameters involved in COVID-19.

Various researches have been undertaken in the field of survival analysis on COVID-19, based on pre-existing patient health conditions. Using accurate patient data, we will be able to estimate hazard rates, lengths of duration, of hospital stay, and ventilator support. These help to evaluate a person's risk of infection. Due to the slow rollout of vaccinations and exponential rate of increase in the number of cases, there is an immediate need for better and more robust methods for tackling the virus.

This chapter will analyze various means of a faster method of testing using chest X-ray and CT scan images on deep neural networks for predicting whether the person is infected or not. Using these methods in conjunction with patient characteristics or pre-existing conditions can improve the performance and reliability of the models. This chapter makes use of various machine learning algorithms to provide future forecasts about the spread of the infection. These are usually regression algorithms which use data from previous months and predict future numbers, and their correctness can be verified using the actual data. Data analysis helps to understand the exposure of various age groups to infection when the factors of lockdown and social distancing are taken into account. Classical time series modeling can be used for important short-term predictions. Various results have been proposed which provide short-term real-time forecasts and risk assessment. This chapter will provide results and comparisons made between various predictive models such as Auto-Regressive Integrated Moving Average (ARIMA), Prophet, and LSTM (Long Short Term Memory).

Compartmental models like SIR and SEIR are important tools in data analysis of an entire population during an epidemic. These models provide the number of days it takes for a disease to reach peak levels of infection.

This chapter has been divided into eight sections. This first section provides the introduction and requirement for data analysis on COVID-19. Section 12.2 provides the literature study which analyzes previous works in the field of data analytics for COVID-19. Section 12.3 gives an exploratory data analysis of the pandemic through graphs and time series data. Section 12.4 introduces compartmental models for epidemiology and will highlight their importance in understanding the spread of the disease in a population. Section 12.5 introduces deep learning techniques for a rapid and accurate diagnosis of COVID-19 in a patient using chest X-ray and chest CT scan images. Section 12.6 introduces survival analysis techniques to understand the probability of survival for a patient infected with COVID-19 based on previous or pre-existing health conditions. Section 12.7 introduces various forecasting models to predict the number of COVID-19 cases on a specific day based on the previous data. Section 12.8 provides the conclusion to the chapter by summarizing all the models which have been introduced in the chapter.

12.2 LITERATURE STUDY

This section will give an insight into the studies conducted in response to the COVID-19 pandemic. Many studies have been published which confirm the recent trends observed in the spread and modeling of the pandemic. From a medical point of view, the novel coronavirus has more in common with SARS-CoV than with MERS-CoV (Kannan et al. 2020; Kundapur et al. 2020; Shereen et al. 2020). Epidemiological models based on compartmental models like SEIR have been studied (Yang et al. 2020; Roda et al. 2020). These papers have attempted to demonstrate the relationship between various measures to control the flow of the virus such as large-scale quarantine, mass lockdown, social distancing, strict regulations on travel, and closure of crowded public spaces. Many clinical studies have been performed on various patient datasets, including an analysis of symptoms and patient medical history (Huang et al. 2020; Mo et al. 2020; Fu et al. 2020). These studies perform deep analysis on the relationship between various symptoms, how their occurrence might be an important indicator for COVID-19, and the relationship between the severity of the infection in the patient with clinical characteristics and the survival probability of the patient (Di Castelnuovo et al. 2020; Li et al. 2020). These studies help to gain an insight into the clinical features a person might show when they are infected.

If the clinical characteristics demonstrate the virus has just started to accumulate, the doctors might suggest lighter treatments, but only based on the person's medical history. If the characteristics suggest that the virus has developed lesions in most parts of the lung, the doctor might recommend the use of high-end treatment options like the use of ventilators which are usually reserved for late-stage patients.

Many papers have been published which make use of deep learning techniques for fast and accurate classification and segmentation of chest X-ray images, CT scan images, or chest radiography images (Jaiswal et al. 2020; Apostolopoulos and Mpesiana 2020; G. Wang et al. 2020; Shi et al. 2020). These techniques have given very promising results which can be used successfully for a faster and more accurate prediction of the presence of COVID-19 in a patient. Segmentation helps us to map the region of infection in the lungs and suggest possible treatments. These methods can be combined with the patient's medical history to make them even more reliable so that they can be accepted into the real-world diagnosis of the patients.

Many studies have attempted to develop forecasting models for COVID-19 using previous infection data (Anastassopoulou et al. 2020; Grasselli et al. 2020; Petropoulos and Makridakis 2020). Few studies have shown good predictive models, although these models are not very reliable because they cannot predict government policies and, most importantly, these models cannot predict if the citizens of a country will act responsibly and follow the rules, like wearing masks and not going out unless necessary. These models have used very basic machine learning techniques to forecast future numbers.

12.3 EXPLORATORY DATA ANALYSIS

This section will provide basic visualizations for the COVID-19 pandemic. These graphs will demonstrate how fast the disease is spreading and convey the severity of the disease. This consists of worldwide number of cases, deaths, and recoveries. A comparative analysis of the countries with highest number of cases will be explored. Specific data for India will be visualized using bar graphs. Figure 12.1a–c will illustrate the number of confirmed cases, confirmed deaths, and confirmed recoveries from COVID-19 in the countries with the four highest number of cases, i.e., the United States, India, Brazil, and Russia. Figure 12.1d will illustrate the total number of confirmed cases in the top ten countries compared against the rest of the cases in the world.

Figure 12.2a–c illustrates the number of confirmed cases in India, the daily increase in the number of cases, and the daily increase in confirmed deaths, along with a seven-day moving average. These figures demonstrate the seriousness of the disease and also the fact that it will be very difficult to cause a descent or even flatten the curve given the rate of increase of infections. As fewer than 1% of the total population has been affected, this demonstrates that if the people do not take proper precautions on their own, every person in the country will be affected.

There are several countries which have had to face a second wave of the virus after they were successful in controlling its spread. Two of these countries include Spain and France, and Figure 12.3a–d illustrates the total number of confirmed cases per day and daily increase in the total number of cases for these two countries. They

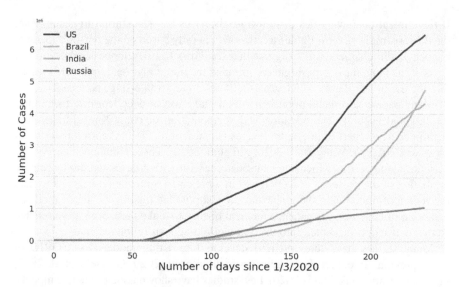

FIGURE 12.1A Number of confirmed COVID-19 cases in the top four countries by case numbers.

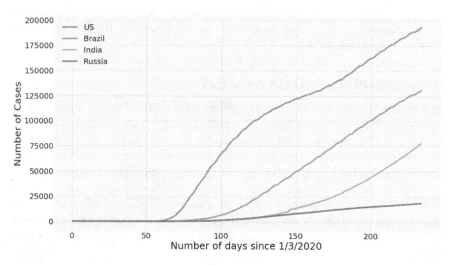

FIGURE 12.1B Number of confirmed COVID-19 deaths in the top four countries by case numbers.

describe the spread of COVID-19 which was originally thought to be controlled but resurfaced again.

The following sections will provide helpful insights into how data analysis can help to forecast future predictions, importance of compartmental models to determine the peak rate of infection, how medical imaging datasets can be helpful in faster and accurate detection of COVID-19, and how survival analysis can determine the probability of survival of a patient infected with COVID-19.

Data Analytics for COVID-19　　　　　　　　　　　　　　　　　　　　　　235

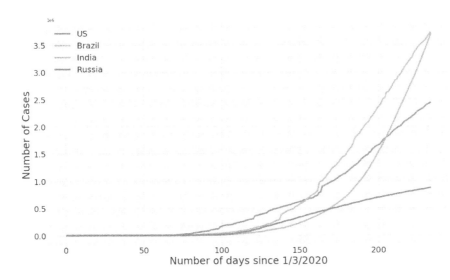

FIGURE 12.1C Number of confirmed COVID-19 recoveries in the top four countries by case numbers.

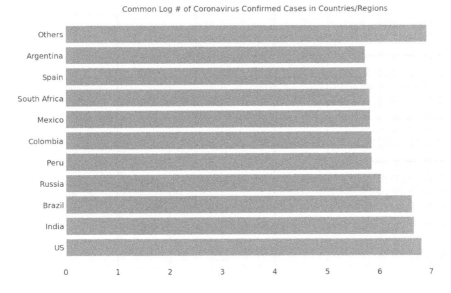

FIGURE 12.1D Total number of confirmed cases in the top ten countries compared against the rest of the cases in the world.

12.4　COMPARTMENTAL MODELS FOR EPIDEMIOLOGY

This section will introduce a few of the compartmental models which simplify the mathematical modeling of infectious diseases. This is done by approximating the total population into different compartments, such as S (Susceptible), I (Infected), E (Exposed), and R (Recovered). The models can be computed using simple

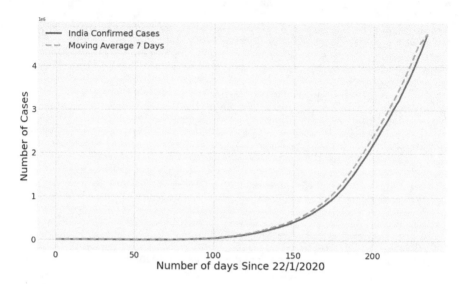

FIGURE 12.2A Total number of confirmed COVID-19 cases in India.

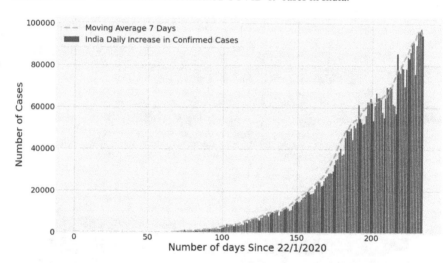

FIGURE 12.2B Daily increase in confirmed cases in India.

mathematical differential equations. Various modifications can be done to the basic SIR model and the compartment labels determine the flow patterns.

Figure 12.4a illustrates the flow of the SIR model. Figure 12.4b illustrates the equations used for determining the infection's spread over time for SIR and SEIR modeling. This can be done using a simple method assuming no population change during a specified time or a dynamic method where the population changes according to a specified rate. This chapter performs simple compartmental model-based SIR modeling on a real-time global COVID-19 dataset obtained from the open

Data Analytics for COVID-19

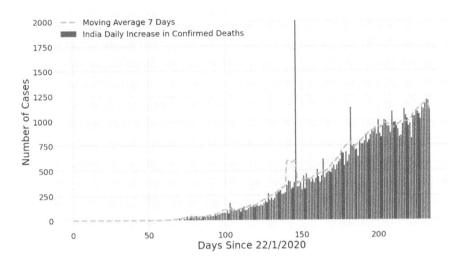

FIGURE 12.2C Daily increase in confirmed deaths in India.

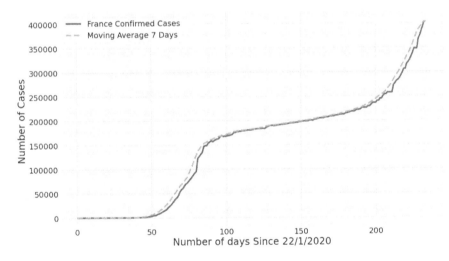

FIGURE 12.3A Total number of confirmed cases in France.

source GitHub repository (Ferlay et al. 2019). The modeling will demonstrate the importance of imposing lockdown. Figure 12.4c illustrates the simple SIR-based modeling extended on the COVID-19 data. This analysis has been performed on the data gathered until August, 2020, which is why the number of case counts does not match the actual numbers. But the model can give a strong perception of the susceptibility, infection, and recovery rates. These models can be accurate for countries with smaller populations, but for countries like India, where even if the number of people infected is large, that is a small percentage of the total population, this model may not produce reliable results.

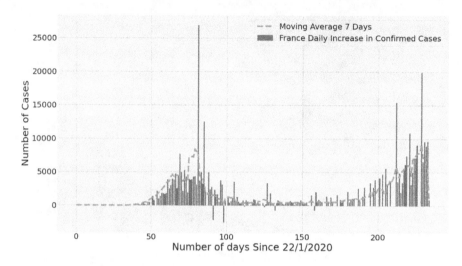

FIGURE 12.3B Daily increase in confirmed cases in France.

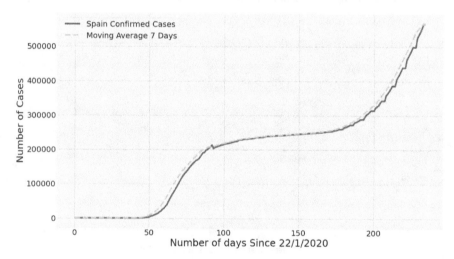

FIGURE 12.3C Total number of confirmed cases in Spain.

Figure 12.5 demonstrates a sharp increase in the number of cases during the start of lockdown 3.0. After the initialization of Unlock 1.0, the number of confirmed cases again had a spike. This meant people had more social contact, less physical distancing, and oversaw increased flouting of the lockdown rules, which directly meant more people becoming infected.

12.5 DEEP LEARNING FOR COVID-19 DIAGNOSIS

Deep learning techniques are the most important area of research in today's world. If anything generates data, machine learning and deep learning technologies can be

Data Analytics for COVID-19

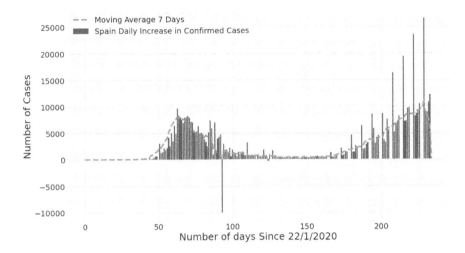

FIGURE 12.3D Daily increase in confirmed cases in Spain.

FIGURE 12.4A SIR model flow.

$$\frac{dS}{dt} = -\frac{\beta IS}{N},$$

$$\frac{dI}{dt} = \frac{\beta IS}{N} - \gamma I,$$

$$\frac{dR}{dt} = \gamma I,$$

FIGURE 12.4B SIR model equations.

applied to predict the results for unknown values. These technologies have immense use in the fast and accurate diagnosis of COVID-19 patients. Chest X-ray images are not very frequently analyzed by clinicians and doctors to find the presence of COVID-19 in patients. Many recent studies have confirmed that the chest X-ray images in a patient identified as COVID-19 positive have specific abnormalities distinct from the chest X-ray of a person with no infection in the lungs (Ng et al. 2020;

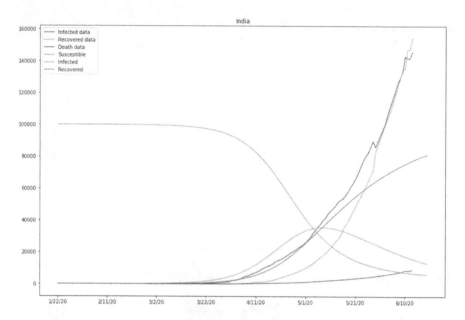

FIGURE 12.4C SIR modeling of the pandemic in India.

Jacobi et al. 2020). Conventional methods of testing are not always very accurate and provide a large number of false-positive and false-negative results. The aim of using deep learning technologies is to identify alternative methods of testing which can prevent physical contact with medical professionals, thus saving much required time which the conventional methods of testing would have taken. This will help in the prevention of the collapse of the health institutions of the country.

This led to the development of deep learning technologies for the accurate classification of chest X-ray images. Various studies have proposed different types of models. Many studies have also proposed transfer learning-based models. Transfer learning is a deep learning technique where the model is not trained on its own. Instead, weights of a different classification task are used as the starting point in these models. They have proved to be very impressive in terms of accuracy and speed. This is because, since the weights are not randomly initialized, the training process takes very little time to reduce the cost function and find the optimum value of accuracy.

Medical image segmentation can be defined as mapping out the region of interest in any kind of X-ray, CT scan, or radiography image. Usually in these images, the infected part of the lung needs to be segmented or pointed out. This will enable the doctor to suggest treatment options based on the severity and spread of the virus lesion. Using deep learning techniques for medical segmentation in CT scans or X-ray images of COVID-19 positive patients, the infected portion of the lung can be accurately mapped out. This automated technique is very powerful and has achieved promising results in terms of the dice similarity coefficient value. This is a value which is a measure of the similarity between two images when the two images are

Data Analytics for COVID-19

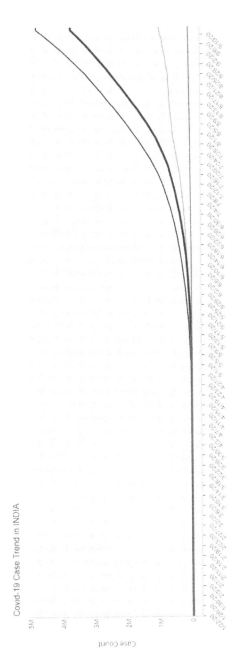

FIGURE 12.5 Importance of social distancing.

TABLE 12.1
Comparison of Accuracies of Different Classification Models

Classification and Prediction Models	Accuracy
Apostolopoulos and Mpesiana 2020	96.78%
Minaee et al. 2020	96%
L. Wang and Wong 2020	93.3%
Jaiswal et al. 2020	96.25%
Maghdid et al. 2020	98%
Razzak et al. 2020	98.75%

TABLE 12.2
Comparison of Dice Similarity Coefficients of Different Segmentation Models

Segmentation Model	Dice Similarity Coefficient
Ahmed et al. 2020	99.26%
Shan et al. 2020	91.6%
Shan et al. 2020	94%
Mishra 2020	97.68%

convolved on each other. The predicted masked region was almost identical to the actual mask region. This method can be helpful in enforcing the practice of physical distancing between the patient and the doctor. Software can be developed with the trained model as its backbone. This software can automatically segment the region of infection from the medical image and send it to the doctor via electronic mail. The doctor can remotely access it and convey the treatment options to the patient without physically ever being in contact with the patient.

Table 12.1 shows the accuracy values achieved in many successful studies which classified chest X-ray and CT scan images of patients to accurately determine the presence of COVID-19 in a patient. Table 12.2 shows the dice similarity coefficient values in many hugely cited studies which performed very accurate segmentation tasks. These techniques give a far-reaching advantage over conventional methods and should be used in the diagnosis process for COVID-19 patients. The study conducted by Apostolopoulos and Mpesiana (2020) utilized the concept of transfer learning by applying it to various individual architectures. The best accuracy was achieved using the MobileNetV2 model. Transfer learning and K-Fold model selection techniques were applied to train the model by Razzak et al. (2020), which achieved 98.75% accuracy. Transfer learning techniques have also been applied by Minaee et al. (2020) and Maghdid et al. (2020). The DenseNet-201 model was used for the classification purpose by Jaiswal et al. (2020). The study conducted by

Data Analytics for COVID-19   243

L. Wang and Wong (2020) made use of contraction and expansion paths for the classification task and compared their model with some basic architectures like VGG16 and ResNet and achieved higher accuracy. The segmentation models use modified forms of the U-Net segmentation model. This can be thought of as the basic model for segmentation which has received widespread positive reviews and has achieved very good results in medical imaging research.

As shown in Tables 12.1 and 12.2, deep learning techniques for medical image classification and segmentation have achieved very promising results and have shown to be highly accurate, reliable, safe, and fast. These techniques need to be quickly put in place by enacting the required legislation so as to reduce the burden on the already crumbling health institutions of the country. Many studies have been conducted which have taken into account the clinical characteristics of the person (Mei et al. 2020). These characteristics include temperature, white blood cell count, neutrophil count, and lymphocyte count, along with the number of patients having symptoms like fever and cough, and finally including their exposure history. These data are essential for the model's reliability in its use in actual medical diagnosis.

12.6 SURVIVAL ANALYSIS FOR COVID-19

Survival analysis is a branch of statistics which is used to determine the expected duration of time until an event occurs using various different algorithms. These events include but are not limited to the death of a person, customer churn and relations, and fault prediction in machines. In the field of COVID-19, or in general medical research, a widely used method is the Kaplan–Meier Curve. It can be used to measure the probability of patients who live for a certain amount of time after treatment. Figure 12.6 illustrates the Kaplan–Meier estimator for the survival function $S(t)$. This returns the probability that life is longer than t. In this equation, t_i is the time when at least one event happened, d_i is the number of events (e.g., deaths) that occurred at time t_i, and n_i is the individuals who are known to have survived till time t_i. This chapter will present the parameterized survival probabilities of COVID-19 patients with respect to their ages. The parameters taken here are the clinical parameters which specify the pre-existing patient health conditions.

The dataset which has been used has been taken from the Mexican Government Open Database ("Datos Abiertos – Dirección General de Epidemiología I Secretaría de Salud I Gobierno I Gob.Mx" 2020). This has been done because of the lack of any good open source datasets from any trusted source. This was the only dataset which was found to have all the required parameters for a good analysis for patient survival as well as for having a large database. This dataset contains more than

$$\widehat{S}(t) = \prod_{i:\ t_i \leq t} \left(1 - \frac{d_i}{n_i}\right)$$

FIGURE 12.6 Kaplan–Meier estimator function.

530,000 patients infected with COVID-19, their age, gender, and pre-existing health conditions such as pneumonia, immunosuppression, and hypertension, along with information about personal choices such as smoking. This dataset included a column which stated whether the patients lived after they were admitted or received treatment at the hospital. Kaplan–Meier function takes two parameters for initialization: the event, which in this case is death, and the tenure, which in this case is the age of the patient. Many survival analysis curves which have been provided give an idea about the survival probabilities of the different patients based on their ages, along with survival probability curves of patients who used to suffer from any pre-existing condition. The analyses between various clinical factors are then compared. Figure 12.7a illustrates the probability of survival of a patient based on age without taking into consideration any parameters. This confirms the general hypothesis that elderly people have a higher probability of dying from the virus than the younger populace. It should be noted that the y-axis represents the probabilities of dying from the virus and the x-axis represents the ages of the population. Figure 12.7b illustrates the comparison of survival probabilities of men and women after contracting the virus with respect to their ages. According to this figure, women have a lower chance of survival than men.

Figure 12.8a gives the comparative survival analysis of patients who had previouly suffered from pneumonia. Figure 12.8b gives the comparative survival analysis of patients who used to smoke tobacco products. According to this curve, only people above 80 years of age have a small difference in survival in comparison to the younger population. Figure 12.8c gives the comparative survival analysis of patients who have obesity. Figure 12.8d–f gives the comparative survival analysis of patients with hypertension, heart problems, and immunosuppression problems. Generally speaking, patients with pre-existing health issues were more susceptible to death than patients who had good health.

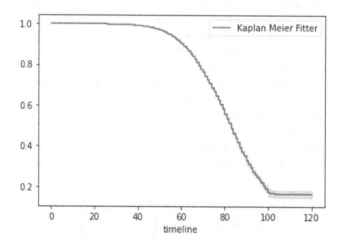

FIGURE 12.7A Basic survival analysis curve.

Data Analytics for COVID-19

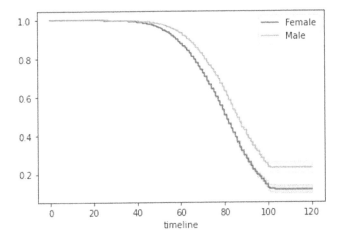

FIGURE 12.7B Comparison of survival analysis by gender.

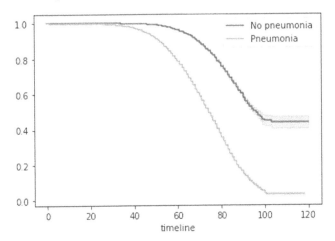

FIGURE 12.8A Comparison of parameterized survival probabilities with age (pneumonia).

12.7 COVID-19 FORECASTING MODELS

Various machine learning algorithms have been developed to predict the number of cases in the near future. These models are developed only to prepare for the upcoming future in terms of medical readiness, treatment, and fast decision-making. These models are not an indication of the exact number of people who are going to be infected but an approximation based on past data. Their accuracy and reliabilities are heavily dependent on country-specific factors such as lockdown rules, policing, important events, big festivals, and most importantly the mindset of the people. The datasets used for this purpose have been obtained from the COVID19-India API database ("COVID19-India API I Api" 2020).

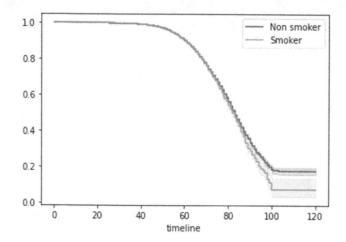

FIGURE 12.8B Comparison of parameterized survival probabilities with age (smoking).

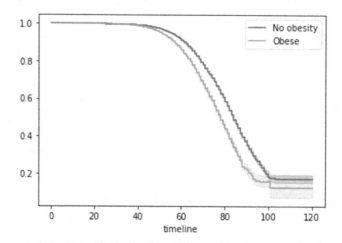

FIGURE 12.8C Comparison of parameterized survival probabilities with age (obesity).

A small implementation of the ARIMA (Auto-Regressive Integrated Moving Average) model to predict the number of cases for the next two days is illustrated in Figure 12.9a. A brief glance at the number of confirmed cases graph gives us an idea of the general pattern being followed. There is a sharp spike in cases followed by a very brief plateau, followed by a sharp decline, which is followed by a spike to a value higher than the previous spike. This is a very general approximation of the pattern of confirmed cases in India. Figure 12.9b shows the forecast with upper and lower bound confidence levels. Figure 12.9c shows a comparison of the predicted values with the original values for a brief period of time. The Prophet model by Facebook also produced very good predictions. This model gave us the trend as well as the weekly trend results.

Data Analytics for COVID-19

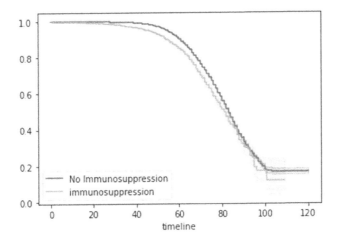

FIGURE 12.8D Comparison of parameterized survival probabilities with age (immunosuppression).

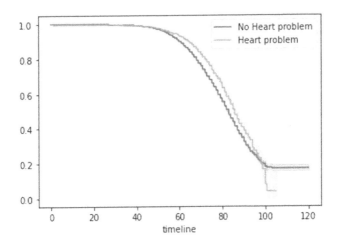

FIGURE 12.8E Comparison of parameterized survival probabilities with age (heart problem).

Figure 12.10 shows the predictions made by the Prophet model. The different types of LSTM models, which are used for anomaly detection and stock prediction, were also analyzed. A random date was chosen, which was "19/06/2020," and the prediction of the number of cases for the next day in India was calculated using each of these models. The ARIMA model gave the output value closest to the actual value. Table 12.3 shows the predicted cases for the next day using different algorithms.

Real-time forecasts and data-driven analysis have proved to be very effective in short-term risk assessment, disease management as well as remaining equipped for future outbreaks (Pandey et al. 2020; Chakraborty and Ghosh 2020). Equipped with

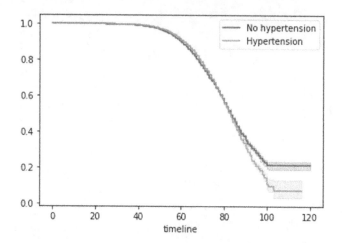

FIGURE 12.8F Comparison of parameterized survival probabilities with age (hypertension).

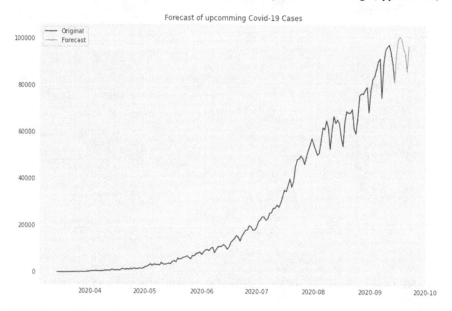

FIGURE 12.9A Future forecasting using the ARIMA model.

the proper dataset, this can help in understanding which section or demographic COVID-19 has affected the most and send aid to the ones in need.

12.8 CONCLUSION

The aim of this chapter is to introduce various data analysis techniques which can prove to be useful in the diagnosis, prediction, classification, segmentation, and forecast of future values for COVID-19. This chapter has given some basic exploratory

Data Analytics for COVID-19

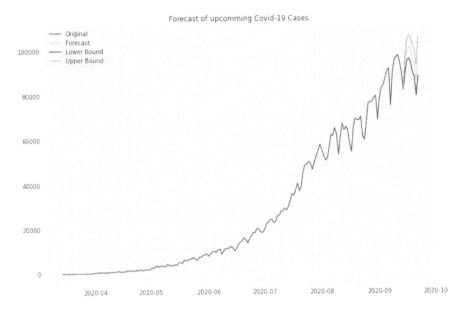

FIGURE 12.9B Upper and lower bound confidence in forecasting.

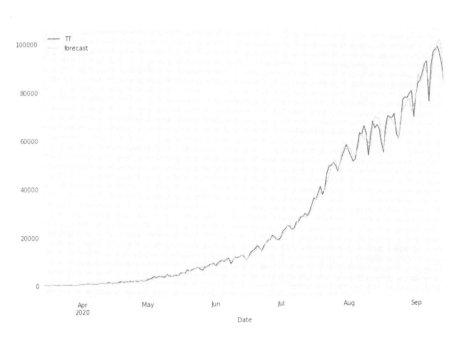

FIGURE 12.9C Original vs predicted values using ARIMA model.

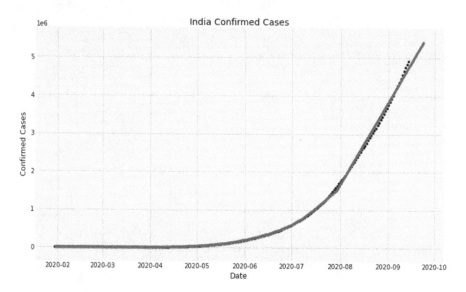

FIGURE 12.10 Future forecasting by the Prophet model.

TABLE 12.3
Comparison of Predicted Cases for 20 June 2020

Model	Prediction for 20/06/20
ARIMA	384,634
Facebook Prophet	360,048
Vanilla LSTM	390,479
Stacked LSTM	396,311
Bidirectional LSTM	393,300
CNN LSTM	384,716
DNN	388,073

analyses on COVID-19 time series data for visualizing the spread of infection. This helps to have an insight into the speed of transmission of the virus. They also convey the fact that COVID-19 has a low mortality rate and a high recovery rate. But this does not stop the virus from damaging economy, global finance, world trade, and tourism opportunities. This pandemic has forced a lot of businesses to the brink of bankruptcy, thus resulting in millions of job losses worldwide. This has also forced the educational institutions to close down and crumble the health institutions of many countries. The probability of a person recovering from the virus is high only as long as the person is treated immediately on contacting the virus, which leads to an urgent need for faster and more accurate testing methods apart from conventional techniques. They include deep learning techniques for automated classification and segmentation of patients based on their chest X-ray images and CT scan images. Many studies have proposed promising results that make use of neural network and transfer learning techniques, which if used in conjunction with patient medical data, can in the near future replace the conventional methods. Studies have been performed on clinical data consisting of the count of white blood cells, temperature of the patient, and other medical tests conducted, also taking into account the chest X-ray image of the patient. This is a form of ensemble learning where the clinical data has to be classified using machine learning techniques apart from the chest X-ray image which is fed as an input to a neural network for feature extraction. Compartmental models for epidemiology such as the SIR model along with forecasting models like ARIMA and Prophet have been analyzed, which can predict the peak of infection, predict the number of cases and deaths in the near future, and help to remain prepared for more patients. Predicting the probability of death of a patient using survival analysis has been introduced and discussed. This analyzes one output event and computes either parameterized or non-parameterized probability of whether that event, which in this case is the death of the patient, will occur with respect to a tenure, which in this case is age. This makes use of pre-existing health conditions of the patient.

REFERENCES

Ahmed, Sifat, Tonmoy Hossain, Oishee Hoque, Sujan Sarker, Sejuti Rahman, and Faisal Shah. 2020. *Automated COVID-19 Detection from Chest X-Ray Images: A High Resolution Network (HRNet) Approach.* doi:10.1101/2020.08.26.20182311.

Anastassopoulou, Cleo, Lucia Russo, Athanasios Tsakris, and Constantinos Siettos. 2020. "Data-Based Analysis, Modelling and Forecasting of the COVID-19 Outbreak." *PLoS ONE* 15 (3). doi:10.1371/journal.pone.0230405.

Apostolopoulos, Ioannis D., and Tzani A. Mpesiana. 2020. "Covid-19: Automatic Detection from X-Ray Images Utilizing Transfer Learning with Convolutional Neural Networks." *Physical and Engineering Sciences in Medicine* 43 (2). doi:10.1007/s13 246-020-00865-4.

Chakraborty, Tanujit, and Indrajit Ghosh. 2020. "Real-Time Forecasts and Risk Assessment of Novel Coronavirus (COVID-19) Cases: A Data-Driven Analysis." *Chaos, Solitons, and Fractals* 135. doi:10.1016/j.chaos.2020.109850.

"COVID19-India API I Api." 2020. https://api.covid19india.org/ Accessed October 23.

"Datos Abiertos – Dirección General de Epidemiología I Secretaría de Salud I Gobierno I Gob. Mx." 2020. https://www.gob.mx/salud/documentos/datos-abiertos-152127 Accessed October 23.

Di Castelnuovo, Augusto, Marialaura Bonaccio, Simona Costanzo, Alessandro Gialluisi, Andrea Antinori, Nausicaa Berselli, Lorenzo Blandi, et al. 2020. "Common Cardiovascular Risk Factors and In-Hospital Mortality in 3,894 Patients with COVID-19: Survival Analysis and Machine Learning-Based Findings from the Multicentre Italian CORIST Study." *Nutrition, Metabolism, and Cardiovascular Diseases.* doi:10.1016/j.numecd.2020.07.031.

Ferlay, J., M. Colombet, I. Soerjomataram, C. Mathers, D. M. Parkin, M. Piñeros, A. Znaor, and F. Bray. 2019. "Estimating the Global Cancer Incidence and Mortality in 2018: GLOBOCAN Sources and Methods." *International Journal of Cancer.* doi:10.1002/ijc.31937.

Fu, Leiwen, Bingyi Wang, Tanwei Yuan, Xiaoting Chen, Yunlong Ao, Thomas Fitzpatrick, Peiyang Li, et al. 2020. "Clinical Characteristics of Coronavirus Disease 2019 (COVID-19) in China: A Systematic Review and Meta-Analysis." *Journal of Infection* 80 (6). doi:10.1016/j.jinf.2020.03.041.

Grasselli, Giacomo, Antonio Pesenti, and Maurizio Cecconi. 2020. "Critical Care Utilization for the COVID-19 Outbreak in Lombardy, Italy: Early Experience and Forecast during an Emergency Response." *JAMA: the Journal of the American Medical Association.* doi:10.1001/jama.2020.4031.

Huang, Chaolin, Yeming Wang, Xingwang Li, Lili Ren, Jianping Zhao, Yi Hu, Li Zhang, et al. 2020. "Clinical Features of Patients Infected with 2019 Novel Coronavirus in Wuhan, China." *The Lancet* 395 (10223). doi:10.1016/S0140-6736(20)30183-5.

Jacobi, Adam, Michael Chung, Adam Bernheim, and Corey Eber. 2020. "Portable Chest X-Ray in Coronavirus Disease-19 (COVID-19): A Pictorial Review." *Clinical Imaging.* doi:10.1016/j.clinimag.2020.04.001.

Jaiswal, Aayush, Neha Gianchandani, Dilbag Singh, Vijay Kumar, and Manjit Kaur. 2020. "Classification of the COVID-19 Infected Patients Using DenseNet201 Based Deep Transfer Learning." *Journal of Biomolecular Structure and Dynamics.* doi:10.1080/07391102.2020.1788642.

Kannan, S., P. Shaik Syed Ali, A. Sheeza, and K. Hemalatha. 2020. "COVID-19 (Novel Coronavirus 2019) - Recent Trends." *European Review for Medical and Pharmacological Sciences* 24 (4). doi:10.26355/eurrev_202002_20378.

Kundapur, Rashmi, Anusha Rashmi, M. Sachin, Karishma Falia, R. A. Remiza, and Shambhavi Bharadwaj. 2020. "Covid 19 – Observations and Speculations – a Trend Analysis." *Indian Journal of Community Health* 32 (2 Special Issue).

Li, Xiaochen, Shuyun Xu, Muqing Yu, Ke Wang, Yu Tao, Ying Zhou, Jing Shi, et al. 2020. "Risk Factors for Severity and Mortality in Adult COVID-19 Inpatients in Wuhan." *Journal of Allergy and Clinical Immunology* 146 (1). doi:10.1016/j.jaci.2020.04.006.

Maghdid, Halgurd, Aras Asaad, Kayhan Ghafoor, Ali Sadiq, and Khurram Khan. 2020. *Diagnosing COVID-19 Pneumonia from X-Ray and CT Images Using Deep Learning and Transfer Learning Algorithms.*

Mei, Xueyan, Hao Chih Lee, Kai yue Diao, Mingqian Huang, Bin Lin, Chenyu Liu, Zongyu Xie, et al. 2020. "Artificial Intelligence–Enabled Rapid Diagnosis of Patients with COVID-19." *Nature Medicine* 26 (8). doi:10.1038/s41591-020-0931-3.

Minaee, Shervin, Rahele Kafieh, Milan Sonka, Shakib Yazdani, and Ghazaleh Jamalipour Soufi. 2020. "Deep-COVID: Predicting COVID-19 from Chest X-Ray Images Using Deep Transfer Learning." *Medical Image Analysis* 65. doi:10.1016/j.media.2020.101794.

Mishra, Shreyas. 2020. *A Novel Automated Method for COVID-19 Infection and Lung Segmentation Using Deep Neural Networks.*

Mo, Pingzheng, Yuanyuan Xing, Yu Xiao, Liping Deng, Qiu Zhao, Hongling Wang, Yong Xiong, et al. 2020. "Clinical Characteristics of Refractory COVID-19 Pneumonia in Wuhan, China." *Clinical Infectious Diseases: An Official Publication of the Infectious Diseases Society of America.* doi:10.1093/cid/ciaa270.

Ng, Ming-Yen, Elaine Y.P. Lee, Jin Yang, Fangfang Yang, Xia Li, Hongxia Wang, Macy Mei-sze Lui, et al. 2020. "Imaging Profile of the COVID-19 Infection: Radiologic Findings and Literature Review." *Radiology: Cardiothoracic Imaging* 2 (1). doi:10.1148/ryct.2020200034.

Pandey, Gaurav, Poonam Chaudhary, Rajan Gupta, and Saibal Pal. 2020. "SEIR and Regression Model Based COVID-19 Outbreak Predictions in India." *ArXiv Preprint ArXiv:2004.00958.*

Petropoulos, Fotios, and Spyros Makridakis. 2020. "Forecasting the Novel Coronavirus COVID-19." *PLoS ONE* 15 (3). doi:10.1371/journal.pone.0231236.

Razzak, Imran, Saeeda Naz, Arshia Rehman, Ahmed Khan, and Ahmad Zaib. 2020. "Improving Coronavirus (COVID-19) Diagnosis Using Deep Transfer Learning." *MedRxiv*, January, doi:10.1101/2020.04.11.20054643.

Roda, Weston C., Marie B. Varughese, Donglin Han, and Michael Y. Li. 2020. "Why Is It Difficult to Accurately Predict the COVID-19 Epidemic?" *Infectious Disease Modelling* 5. doi:10.1016/j.idm.2020.03.001.

Shan, Fei, Yaozong Gao, Jun Wang, Weiya Shi, Nannan Shi, Miaofei Han, Zhong Xue, Dinggang Shen, and Yuxin Shi. 2020. *Lung Infection Quantification of COVID-19 in CT Images with Deep Learning.*

Shereen, Muhammad Adnan, Suliman Khan, Abeer Kazmi, Nadia Bashir, and Rabeea Siddique. 2020. "COVID-19 Infection: Origin, Transmission, and Characteristics of Human Coronaviruses." *Journal of Advanced Research.* doi:10.1016/j.jare.2020.03.005.

Shi, Feng, Jun Wang, Jun Shi, Ziyan Wu, Qian Wang, Zhenyu Tang, Kelei He, Yinghuan Shi, and Dinggang Shen. 2020. "Review of Artificial Intelligence Techniques in Imaging Data Acquisition, Segmentation and Diagnosis for COVID-19." *IEEE Reviews in Biomedical Engineering.* doi:10.1109/RBME.2020.2987975.

Wang, Guotai, Xinglong Liu, Chaoping Li, Zhiyong Xu, Jiugen Ruan, Haifeng Zhu, Tao Meng, Kang Li, Ning Huang, and Shaoting Zhang. 2020. "A Noise-Robust Framework for Automatic Segmentation of COVID-19 Pneumonia Lesions from CT Images." *IEEE Transactions on Medical Imaging* 39 (8). doi:10.1109/TMI.2020.3000314.

Wang, Linda, and Alexander Wong. 2020. *COVID-Net: A Tailored Deep Convolutional Neural Network Design for Detection of COVID-19 Cases from Chest Radiography Images.*

Yang, Zifeng, Zhiqi Zeng, Ke Wang, Sook San Wong, Wenhua Liang, Mark Zanin, Peng Liu, et al. 2020. "Modified SEIR and AI Prediction of the Epidemics Trend of COVID-19 in China under Public Health Interventions." *Journal of Thoracic Disease* 12 (3). doi:10.21037/jtd.2020.02.64.

Index

A

Acute respiratory distress syndrome (ARDS), 213
Agent-Based Model (AGM), 205
Anonymity, 181
Artificial intelligence (AI), 7, 77
Artificial neural networks (ANN), 218
Autoregressive integrated moving average (ARIMA), 19, 23, 42, 48, 249

B

Blockchain, 1, 11, 16, 17, 66, 70, 71, 73, 91

C

Case fatality rate (CFR), 9
Clustering, 19
Coatomer protein complex (COPB2), 75
Community-acquired pneumonia (CAP), 74
Compartmental model, 234
Computer tomography, 210
Confusion matrix, 223, 224
Convolutional neural networks (CNNs), 32, 33, 42, 218
Corona, 2, 113
COVID-19, 1, 3, 5, 7–9, 11, 12, 17, 23, 24, 29, 32–34, 69, 74, 76, 96, 103, 107, 109–112, 128, 146, 148, 152, 166, 182, 190, 220
COVID-19 intelligent diagnosis and treatment assistant program (NCapp), 148
COVI-SCANNER, 1, 13, 21, 29, 31, 32

D

Database, 219
Data collection, 110
Deep learning, 7, 210
Deep learning models (DLMs), 30, 39
Deep neural network (DNN), 10
Digital drug control system (DDCS), 76
Drone, 51, 52, 81

E

E-commerce, 118
Edge computing, 132
Electronic speed controllers (ESC), 51
Entropy, 23

F

F1 score, 221
False negative, 210
Field of view (FOV), 150
Fog computing, 132

G

Global dense feature (GDF), 37
Ground glass opacity (GGO), 2, 10, 213

H

Healthcare Data Gateway (HGD), 73
Herd immunity, 204
Hierarchical Agglomerative Clustering (HAC), 21
HIV, 76

I

Immunity, 204
International Committee of Taxonomy of Viruses (ICTV), 3
Internet of Things (IoT), 50, 78, 168

J

Java virtual machine (JVM), 111

K

Kaplan Meier function, 244
K-Fold model selection, 242
K-means clustering, 20, 21
K-nearest neighbors (KNN), 10
KNN classifier, 10

L

Learning rate, 218

M

Machine learning, 7, 69
Manta-Ray Foraging Optimization (MRFO), 218
Middle East Respiratory Syndrome Coronavirus (MERS-CoV), 232
MobileNet, 136

Modified Generative Adversarial Network (MGAN), 30, 35, 40, 42
Multimodality, 102
Multiobjective differential evolution (MODE), 218

P

Part of speech (POS), 115
Perception, 180
Personal protective equipment (PPE), 6
Polynomial neural networks (PNN), 71

R

R2 score, 9
Radiofrequency identification (RFID), 1, 69, 177
Remote monitoring, 174
Reverse Transcriptase Polymerase Chain Reaction (RT-PCR), 2, 6, 10, 216
RNA, 2
Robot, 81

S

SARS virus, 3, 18, 29
Sentiment analysis, 114
SequeezeNet, 225
SIR model, 193, 239
Social distancing, 148
Social Internet of Things (SIoT), 131
Social media, 105, 107

Support vector machine (SVM), 8, 9, 211
SUQC model, 193
Susceptible–Exposed–Infected–Asymptomatic–Recovered (SEIAR) model, 197
Susceptible–Exposed–Infected–Recovered (SEIR), 190
Susceptible–Infected–Recovered (SIR) model, 190
Synthetic minority over-sampling technique (SMOTE), 32

T

Text analysis, 114, 117
Text mining, 2, 9, 81, 109
Thermal camera, 150
Thermal imaging, 149
Timestaming, 156
True Negatives Rate (TNR), 41
True positive, 220
True Positive Rate (TPR), 41

U

Unmanned aerial vehicle (UAV), 50, 53
Unsupervised machine learning, 109

W

White lung, 214
World Wide Web (WWW), 96